Grenville Robson

The Analysis of Biological Populations

Special Topics in Biology Series

The Analysis of Biological Populations

Mark Williamson
Professor of Biology, University of York

Edward Arnold

First published 1972
by Edward Arnold (Publishers) Limited,
41 Maddox Street,
London,
W1R OAN

Boards Edition ISBN 07131 2346 X
Paper Edition ISBN 07131 2347 8

Printed in Great Britain by
C. Tinling & Co., Ltd, London & Prescot

Preface

Analysis is one aspect of the science of ecology, at the moment involving a mixture of guess work, skill and experience. Up to now, there has been little such experience, and it is necessary in this book to describe both some ways of analysing populations and the basic theory of populations that underlies any analysis. As analysis has not often been thought of as a distinct aspect of population biology, it is undeveloped and disjointed, and so I can only hope to deal with some aspects of its present state. A book this length has to be selective, and I have not, for instance, considered pattern in communities, partly because I think the study of it tells us rather little about population dynamics, but more because it has been well treated by Pielou (1969). Some interesting theories have not been discussed, as they seem to me not to be relatable as yet to the measurements that can be made in real data. My conviction is that theory, experiments and observations must be developed together, and that this union is the job of the population analyst.

The study of populations is inevitably a quantitative one, and so involves mathematics. In my view, it is impossible to be precise about many of the concepts except by expressing them mathematically. All the same, we know so little about populations that the mathematics needed at the moment is still very simple, and well within the grasp of anyone who can write a simple computer program. I myself prefer using algebra to calculus, for a variety of reasons. The worst of these is that most biologists find it easier. Algebraic formulations can nearly always be solved, while differential equations frequently lead to new forms. For complex systems, matrix representations are both easy to write down and the natural form for computing. It is fairly seldom that we make really continuous observations of populations, which would require calculus: usually we observe in discreet intervals of time, and algebraic representations are sufficient. But when all that is said, there are still plenty of occasions where calculus is clearly better, and so I have not hesitated to include some bits of simple calculus, with an indication of how I think they can be tested. One convenient convention is to use 'log' only for logarithms to the base ten, and to use 'ln' for natural logarithms, logarithms to the base e.

The book is in four sections. The first deals with single populations only, ignoring their genetic aspects. Starting with an account of the observed variability of natural populations, Chapters 2 and 3 deal with some theory and experimental background, while Chapter 4 dicusses the only system for which there is a complete mathematical model, continuous cultures. A model for the exploited phase of fisheries is also outlined in Chapter 4. The section ends with Chapter 5 in which ways of analysing single populations are discussed. The second section deals with some genetic aspects of natural populations. This is a rapidly advancing topic, and only a few aspects are considered, those which have particular relevance to the ecological analysis. The third section deals with the interactions between two species. Two major interactions are discussed: competition and feeding on one species by another. The classification of interactions and some other problems are discussed in Chapter 9. Chapters 10, 11 and 12 consider the theory of competition, experiments on competition and the problem of analysing and detecting competition in natural populations. Predator/prey and other feeding interactions are considered in Chapters 13, 14 and 15. The genetical consequences of these interactions are discussed briefly in Chapters 11 and 15. The final section is concerned with the interactions of many species, or as some would have it, the analysis of communities. The emphasis here is on the quantitative measurements that can be made on systems of species, which may help in the analysis of their interactions and the study of the population densities and the way they change.

As population biology lies between ecology, genetics and evolutionary studies, it is inevitable that some of the material of this book has been presented quite adequately in text books. I have used examples to show what seem to me the main features of the field as we know it at present, without great regard to whether other authors have or have not used these examples. In a few cases I have deliberately reused well known examples, when I have thought biologists could well be more critical and analytic in their approach.

York, 1971 M. W.

Acknowledgments

Much of this book was written on holiday with my family in the Lake District and I am very grateful to my wife, Charlotte, and children for helping me to get on with it. Charlotte has also read the whole book, and made many improvements both grammatical and scientific. Sections I and IV were read by Dr. Michael Usher, Sections II and III by Professor John Currey, and all three readers have done a great deal to make the text clearer and I am very grateful to them. I also thank Dr. David White for comments on various points, particularly in Chapters 3 and 15, and Professor Tom Reynoldson for comments on Chapter 12. The obscurities and mistakes remain my own. I am also grateful to Margaret Eslick for struggling hard with my mumbles on tape and with my writing to produce an adequate typescript, and to Ron Grainger for drawing or sketching many of the figures. Many of the ideas and approaches in this book have been developed from teaching in Edinburgh and York Universities, and I acknowledge the criticism of many students. In the last stage of writing, I was at the Advanced Study Institute on the Dynamics of Numbers in Populations, at Oosterbeek in the Netherlands, and received many suggestions, and I am grateful to all of those who were there. I am also grateful to my Publishers for waiting many years for this book. If you want to write a book do not start a new Department.

I am most grateful to a number of people for permission to copy figures and tables that have been published before. The figures, tables and people involved are:

Fig. 1.1 Drs. H. N. Kluyver and D. Lack and the Delegates of the University Press, Oxford.

Figs. 1.2, 1.3, 1.4 & 1.16 Dr. D. Lack and the Delegates of the University Press, Oxford.

Fig. 1.6 Mr. C. Elton, the British Ecological Society and Blackwell's Scientific Publications Ltd.

Figs. 1.7 and 1.8 Professor P. A. P. Moran, The British Ecological Society, and Blackwell's Scientific Publications Ltd.

Fig. 1.10 The Royal Meteorological Society.

Figs. 1.12 and 1.13 Dr. C. O. Tamm and Munksgaard International Booksellers and Publishers.

Fig. 1.15 Professor G. C. Varley, The British Ecological Society and Blackwell's Scientific Publications Ltd.

Figs. 1.17 and 5.1 Professor Dr. H. Klomp and Academic Press Inc.

Fig. 1.18 Dr. H. G. Vevers, The British Ecological Society and Blackwell's Scientific Publications Ltd.

Fig. 1.19 The U.S. Fish and Wildlife Service.
Figs. 1.20 and 16.3 The International Council for the Exploration of the Sea.
Fig. 2.2 The British Ecological Society and Blackwell's Scientific Publications Ltd.
Fig. 2.3 Mr. M. E. Solomon and the Editors of the Annals of Applied Biology.
Fig. 3.1 John Wiley & Sons Inc., Publishers.
Fig. 4.4 Mr. R. J. H. Beverton and the Controller of Her Majesty's Stationery Office.
Fig. 5.2 Professor T. R. E. Southwood.
Fig. 6.1 The Editors of Heredity.
Table 9.1 Dr. E. B. Broadhead and the Managers of Ecological Monographs.
Fig. 10.2 Professor J. L. Harper and the Syracuse University Press.
Fig. 12.1 Dr. H. G. Lloyd, The British Ecological Society and Blackwell's Scientific Publications.
Fig. 12.3 Dr. T. B. Reynoldson and Academic Press Inc.
Fig. 13.1 Dr. J. M. Colebrook and Mr. G. A. Robinson.
Fig. 14.2 and Table 14.2 Mr. P. H. Leslie, Dr. J. C. Gower and the Trustees and Editors of Biometrika.

York, 1971 M.W.

Contents

IV MANY SPECIES

I The ecology of single species populations

1 Variability in natural populations

By the analysis of a biological population is meant the analysis of why a population is as large as it is, and the analysis of why it changes in size. Before this can be done, it is necessary to know what is usually meant by 'a population', and it is useful to know something about the extent and nature of variation in their size. This Chapter considers a number of populations that have been studied more or less extensively, to illustrate these points.

A population is simply the total number of individuals of a single species in one place. In this book, systems in which more than one species are studied will be regarded as a mixture of several populations: occasionally these have been referred to, in other books, as mixed species populations. Such multiple populations are considered in Sections III and IV. The first two Sections will be concerned only with single populations, in which one species only is studied at a time. The word 'number' comes into the definition of a population, and so the subject is inevitably more or less mathematical. All studies of populations must involve some method of counting. This may be either a total census or it may be based on samples. The total census, counting all the individuals in the population, can only be done with large conspicuous organisms, notably man and some birds. These are the most satisfactory populations statistics, and the early examples in this Chapter will consider them. In other organisms, when samples are taken, one deals with figures such as 'numbers per square metre'. From these estimates, it is of course possible to estimate the total population size, but in such cases the major feature in the figures is the relative change from sample to sample, and not the absolute changes in population size.

There are two questions that can be considered straight away. First, what is meant by 'one place', in the definition given in the paragraph above, and second, how variable are the numbers in real populations?

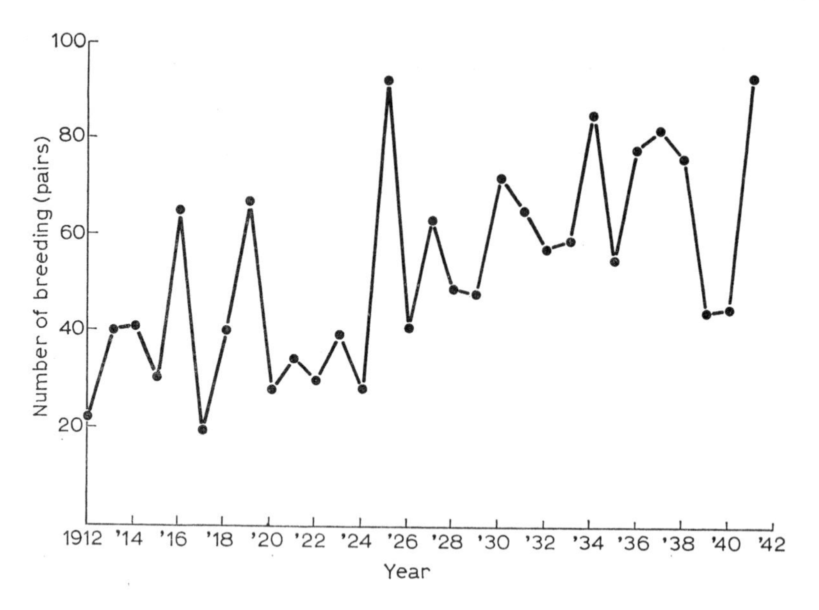

Fig. 1.1 Breeding population of the great tit, *Parus major,* at Oranje Nassau s Oord, near Wageningen, Netherlands (from Kluijver, 1951 and Lack 1966).

Both questions can most easily be answered by illustrating the results of some counts that have been published. Figure 1.1 shows the number of great tits (*Parus major*) in a wood in Holland. The number has varied from about 20 to about 100, or about five times, in a period of thirty years. In this case, one place simply means a wood in which it is convenient to count the great tits. Figure 1.2 shows a similar population, in a wood near Oxford, Marley Wood, which is part of Wytham Woods. This second figure shows the variation within the year as well as the variation from year to year, and shows the effect of counting the young, just hatched, as well as the adult breeding population. The problems of dealing with populations sub-divided in various ways will be considered in more detail both in this Chapter and later in the book, but to simplify the discussion of the variability of the populations all the remaining examples consider only year to year variation. Figure 1.3 shows population changes in a large and conspicuous bird, the heron. Here there seems to be rather little variation in population size, particularly in Cheshire and south Lancashire, except after a 'hard winter'. In the Thames drainage area there appears to have been an increase in population size during the 1950's compared with the earlier period. Again it can be seen that 'one place' is simply an area that is convenient for counting the population that is being studied.

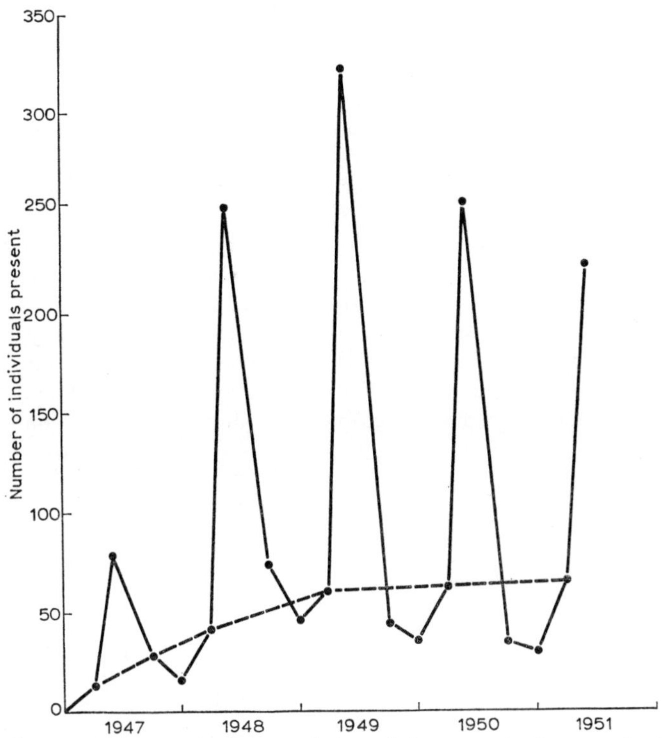

Fig. 1.2 Seasonal changes in the population of the great tit, *Parus major*, in Marley Wood, near Oxford (from Lack 1954 after Gibb 1954).

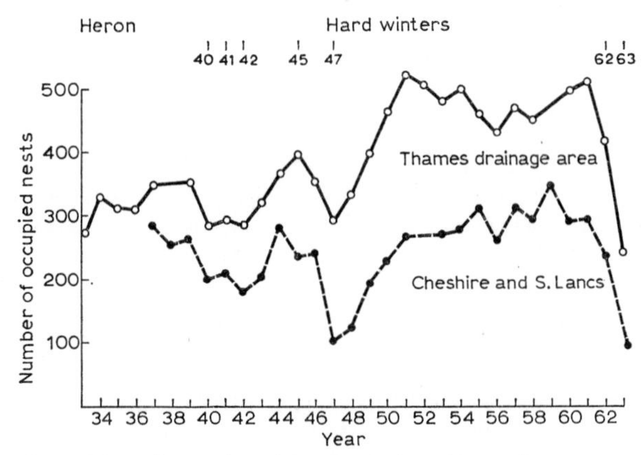

Fig. 1.3 Number of breeding pairs of heron, *Ardea cinerea*, in two parts of England, 1933–1963. Data from the British Trust for Ornithology analysed by J. Stafford (from Lack 1966).

The advantage of logarithmic plots

So far, all the figures have shown the absolute population size, plotted arithmetically, against time. With the rather small variation shown by these populations, there is little advantage in plotting them any other way, but for many populations a logarithmic plot is more helpful and more useful. There are four main reasons for this. The first is that relative changes in populations are most easily studied on logarithmic plots. A doubling of the population size, say, from 10 to 20 or from 100 to 200, produces the same change in a logarithmic plot, but a very different change on an arithmetic plot. In most population analyses we will be concerned with what causes the relative change in population. The second reason is that the logarithm of population size is more symmetrically distributed than the population size itself. This is discussed in detail below (see Fig. 1.9), and becomes more striking the more variable the population is. The third reason is that many populations are so variable, that their variation can only conveni-

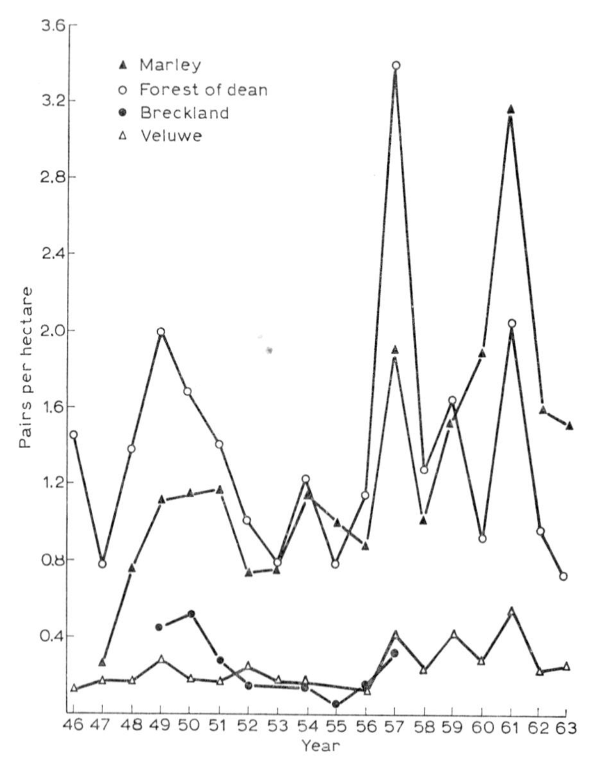

Fig. 1.4 Density of breeding pairs of great tit in Marley Wood, near Oxford, in the Forest of Dean, Gloucestershire, in Breckland, Norfolk, all in England, and in de Hoge Veluwe, Netherlands (from Lack 1966).

ently be shown on a logarithmic plot. An arithmetic plot disguises the variation that takes place, and this again will be shown below. The fourth reason is that the change in the logarithm of the numbers of a population fits in most conveniently with the basic theory of population dynamics, and this point will be discussed in the next Chapter.

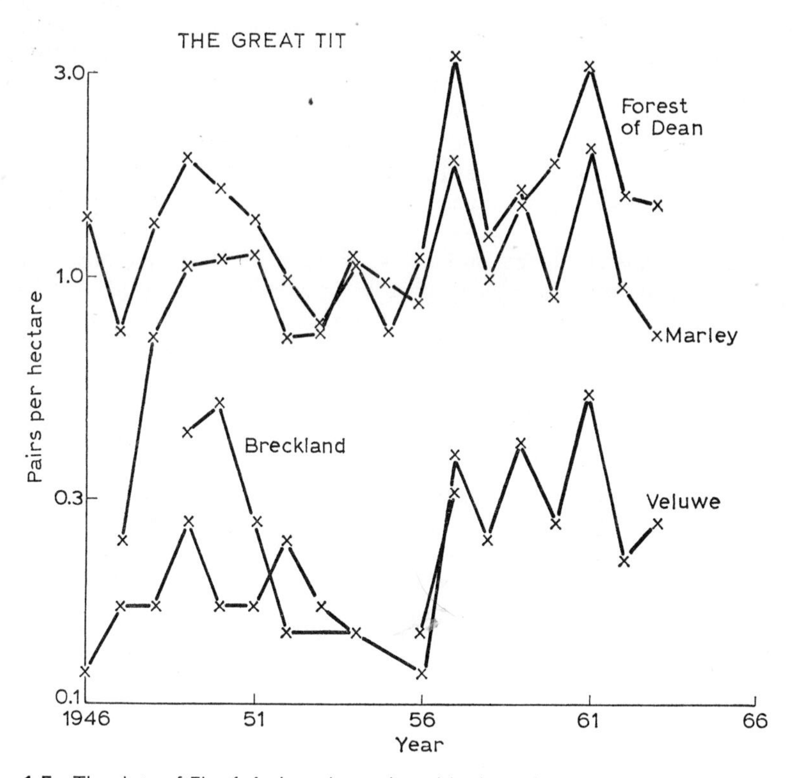

Fig. 1.5 The data of Fig. 1.4 plotted on a logarithmic scale.

Some of these points can be illustrated with populations of the great tit. Figures 1.4 and 1.5 show four populations of these birds plotted arithmetically in Fig. 1.4 and logarithmically in Fig. 1.5. Both figures show the number of pairs per hectare in four different woods, Marley, the wood of Fig. 1.2; Veluwe, the wood of Fig. 1.1; and two other populations in the Forest of Dean in the West of England and Breckland in the East of England. In both figures it is clear that the population density is much higher in Marley and the Forest of Dean than it is in Breckland and Veluwe. The arithmetic plot gives the impression that the commoner populations are more variable. In fact the relative variation is much the same in all four populations as can be seen in the logarithmic plot, Fig. 1.5. The natural statistic for measuring this variation is the standard deviation of the logarithm of the population size. Using logarithms to the base ten, this is 0.23 at Marley, 0.17 at the Forest of Dean, 0.33 Breckland and 0.18 at Veluwe. It is perhaps easier to consider the relative variation that would be expected to be seen in a period of ten years or rather more. This is given by the anti-logarithm of three times this standard deviation. For Marley it is 5.0 times, for Forest of Dean 3.4, for Breckland 9.8 and for Veluwe 3.5. So it seems for any population of the Tits you might reasonably expect the population size to vary between from three times to ten times the minimum observed. This sort of statistic is quite independent of the absolute population size, and so can be useful in comparing populations of many different sorts of organisms. Lewontin (1966) pointed out the value of the standard deviation of logarithmic measurements for this sort of comparison in a rather different context.

The populations considered so far have all been birds, though of course the population of any organism, plant, microbe or animal can be studied. As I am considering only year to year variations, I shall not present any statistics on microbes, but the remaining examples cover a variety of plants and animals. The next one is a fairly large carnivorous animal, the Canadian lynx (*Lynx canadensis*). The number of pelts of this species taken by the Hudson Bay Company from the Mackenzie river district are shown in Figs. 1.6 and 1.7. Figure 1.6 is the arithmetic plot, and gives the impression that the cycles are rather asymmetrical and that there is more variation in the peak numbers than there is in the low

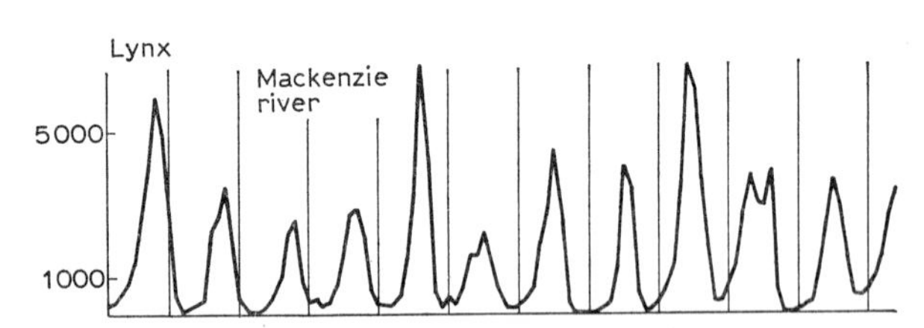

Fig. 1.6 The number of lynx pelts sold by the Hudson's Bay Company from the Mackenzie River District, trapped in the years 1820–1934. (from Elton and Nicholson 1942).

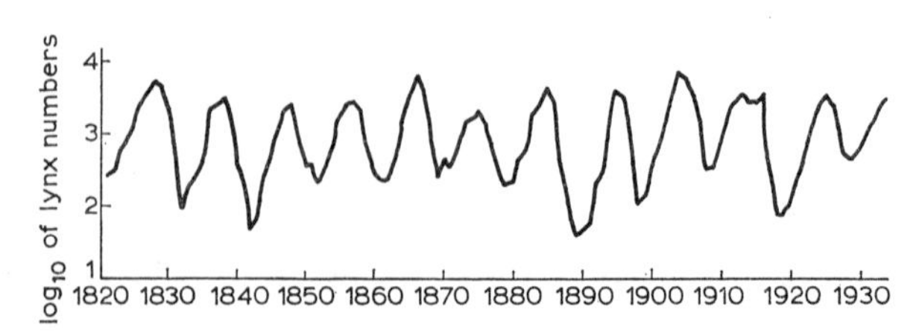

Fig. 1.7 A logarithmic plot of the data of Fig. 1.6 (from Moran 1949).

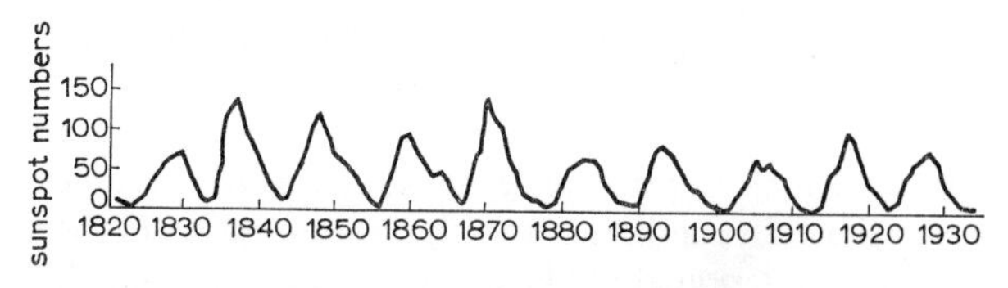

Fig. 1.8 The sunspot cycle (from Moran 1949).

numbers. Figure 1.7 rather contradicts this, showing that the cycles are fairly symmetrical, and that there is in fact more relative variation in the low numbers. Both graphs show very clearly the strongly cyclical variation in the population numbers of this species. This type of cyclical variation is unusual, but not confined to the lynx, or even to the other animals of the Canadian pine forest. For instance there is a clear cycle in the numbers of ptarmigan (Lack 1966), and in the larch bud moth (Baltensweiler 1968) but these clearly marked cycles are distinctly unusual; most populations show the irregular fluctuations we have seen in the earlier figures. Further aspects of the lynx cycle will be considered in Chapter 5.

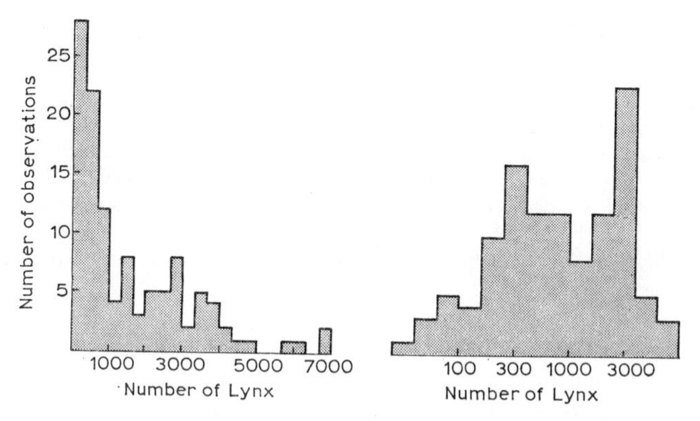

Fig. 1.9 The distribution of numbers of lynx trapped in the Mackenzie River District, considered (a) arithmetically and (b) logarithmically.

There is however one feature which the lynx population shows as clearly as any, and this is shown in Fig. 1.9. This pair of histograms shows the distribution over 114 years both of the arithmetic numbers of lynx and of the logarithmic numbers. It is strikingly clear that the distribution of logarithmic numbers is symmetrical while the other distribution is strongly skewed. This is the usual finding with any extensive population data, and is a very cogent reason for studying the logarithm of population size.

In both the heron and the lynx, a relation has been postulated between physical factors and the change in population size. In the heron this is the hard winters which have been indicated in Fig. 1.3, while with the lynx it is the sunspot cycle, which is shown in Fig. 1.8. In the heron there seems to be a clear relation between the fall of population size and the hard winters that are shown there, but it is not clear there that the hardness of a winter is something that can be measured quantitatively, and is not merely the impression of the ornithologist. If the changes in a population are to be analysed so one can say that a particular factor will affect a population in a predictable way, we need quantitative measures of the factor as well as of the population. Those who lived in England over the period shown in Fig. 1.3 will not be surprised that the hard winters can be quantified, and this is shown in Fig. 1.10 where the four hard winters of 1929, 1940, 1947 and 1963 show up clearly. This figure shows the average temperature, in degrees centigrade, for December, January and February in each year. How-

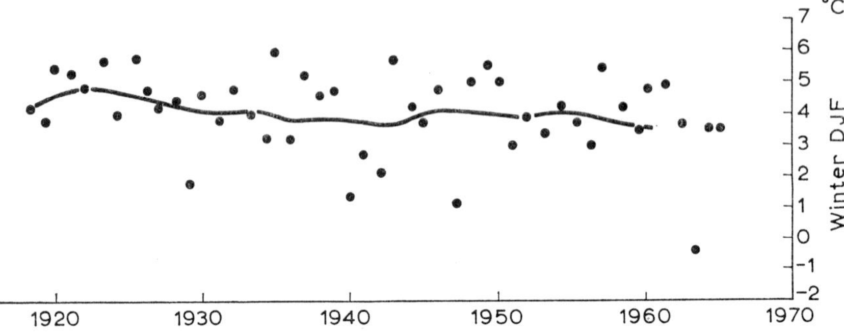

Fig. 1.10 Mean temperature for December, January and February, plotted against the year of January and February, and decade running averages plotted against their mid point, for lowland central England (from Lamb, Lewis and Woodroffe 1966).

ever, nobody seems to have tried to measure the expected fall in the heron population for a particular degree of coldness. With the lynx population, the situation is quite different. Both the sunspots and the lynx population show strong cyclical changes over a century. But the curves were roughly in phase in the 1820's, out of phase in the 1860's, and in phase again in the 20th Century, and so there is clearly no relation between them. This at least shows that some population analyses can be done quickly, if the data are properly presented in a graphical way.

The populations that we have considered so far are only moderately variable. This is shown in Table 1.1 where the standard deviation of the logarithm of population size is shown.*

* This statistic seems not to have been used before, though Watt (1965) used the standard error of the population size. This will of course decrease as the series gets longer, but he was working with series that were all of the same length, and so the difference was not important to him.

Table 1.1 Variability of some populations. For further explanation see text. The numbers against the plankton refer to the lines in Fig. 1.21.

Organism	Time/Stage	Place	No. years	s.d. log pop. size	antilog (3xs.d.)
Mammals					
Lynx		Mackenzie	114	0.558	47.3
Birds					
Great Tit		Forest of Dean	17	0.177	3.39
		Marley	18	0.232	4.97
		Breckland	9	0.331	9.85
		Veluwe	17	0.181	3.50
Yellow-eyed Penguin		New Zealand	16	0.036	1.29
Pied Fly-Catcher		Forest of Dean	16	0.085	1.80
Tawny Owl		Wytham	13	0.085	1.80
Plants					
Anemone hepatica	Plot 2	Sweden	13	0.069	1.61
Anemone hepatica	Plot 3	Sweden	13	0.071	1.63
Insects					
Panaxia dominula		Cothill	29	0.540	41.7
Panolis	Pupae	Letzlinger	60	0.686	114
Hyloicus	Pupae	Letzlinger	49	0.608	66.5
Dendrolimus	Hibeting larvae	Letzlinger	59	1.026	1195
Bupalus	Pupae	Letzlinger	58	1.105	2067
Bupalus	Nymphs	Veluwe	12	0.314	8.74
Bupalus	Pupae	Veluwe	14	0.444	21.4
Plankton of the north-west North Sea (23 entities)					
(23) Calanus finmarchicus stages I-IV (least variance)			11	0.236	5.10
(13) Chaetognaths (median variance)			11	0.492	30.0
(20) Clione limacina (greatest variance)			11	1.011	1081

If the logarithm of the population size followed the normal (or Gaussian) distribution, then about 90% of the observation would fall within one and a half standard deviations of the mean, that is, in a range of three standard deviations. The antilogarithm of this range, that is to say the multiplicative factor from the expected minimum to the expected maximum is also shown in Table 1.1. This range should show the variation that would be expected to occur in about a ten year period, as one observation in ten would fall outside it. In fact, the distribution of the logarithm of the population size is usually slightly flattened at the mean, and drawn in at the tails, platykurtic, and this can be seen in Fig. 1.9 for the lynx. With a platykurtic distribution, a range of three standard deviations covers rather more than 90% of the observations, or the expected variation in a period rather more than ten years, indeed up to twenty years. Using this three deviation measure, the great tit populations vary from three to ten times, the lynx about fifty times. Figure 1.11 shows three less variable bird populations, the pied flycatcher (Lack, 1966), the yellow eyed penguin (Lack, 1966, Richdale 1957) which has a life expectancy of 7 years, and the tawny owl in Wytham Woods, where the variation comes from a steady change in population size rather than irregular oscillations. It is not surprising that longer lived birds show less variable populations but the pied flycatcher is a passerine like the great tit, even though its population is less variable.

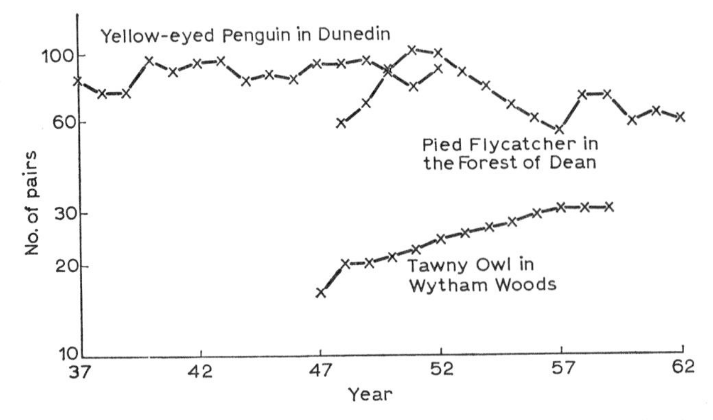

Fig. 1.11 Populations of the yellow-eyed penguin, *Megadyptes antipodes,* the pied flycatcher, *Ficedula hypoleuca,* and the tawny owl, *Strix aluco,* from data quoted by Lack (1966).

Populations of higher plants seem often to be as little variable as those of the longer lived birds, and this may be because they too are long lived. Figures 1.12 and 1.13 show two populations of *Anemone hepatica* studied by Tamm (1956), [see also Harper (1967)]. Each line represents a single plant, and a thicker line shows that the plant flowered. In plot 3 (Fig. 1.12) there is some turnover of the population, with some plants dying and seedlings becoming established, but most of the plants lived the full 13 years of observation. In plot 2 (Fig. 1.13) there was more splitting of individual plants and rather less turnover by death and seedlings.

Populations of invertebrates and algae are usually more variable, partly no doubt because of their shorter life history. Figure 1.14 shows the population of the scarlet tiger moth, *Panaxia dominula* at Cothill near Oxford. It was possible to estimate this population by capture, mark and recapture from 1941 to 1961, the longest period for any organism. The details of estimating a population this way are given in Southwood (1966). This colony of *Panaxia* is almost as variable as the lynx populations as can be seen by Table 1.1. It has another more unusual feature. It occupies an area of marshy ground surrounded by farmland. So in this case the limits of the population are fairly well defined, unlike the other populations that have been considered before and will be considered later in this Chapter. In this case the definition of 'one place' is given by the biology of the

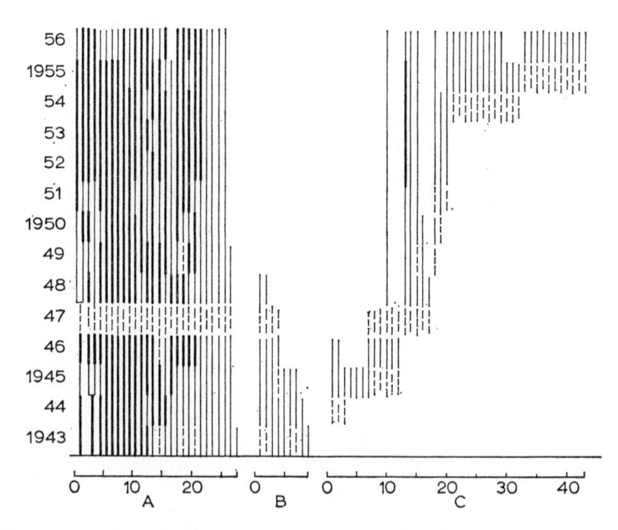

Fig. 1.12 The population of *Anemone hepatica* on plot 3 in a forest in central Sweden. Each line represents one plant, thick when it flowered, thin when it did not, and dashed when it was not seen but is assumed to have been present. When the line divides, the plant has ramified. Group A are the large or intermediate sized plants in 1943, Group B small plants in 1943, Group C appeared later, presumably from seedlings. (from Tamm 1956).

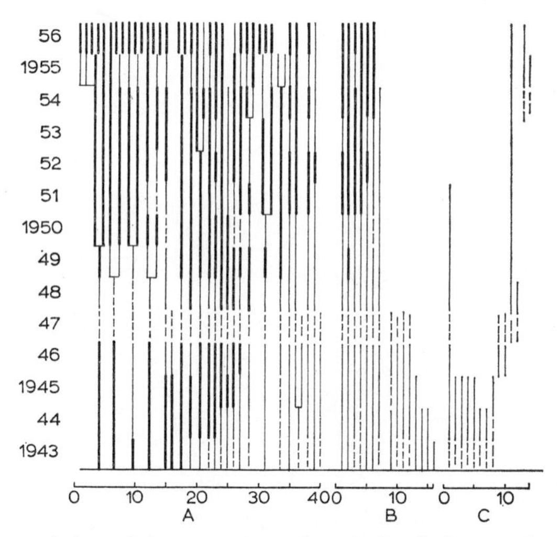

Fig. 1.13 The population of *Anemone hepatica* on plot 2. Conventions as in Fig. 1.12 (from Tamm 1956).

species, and not by the common sense of the observer. This population will be discussed again in Chapter 7.

Four more variable insect populations are shown in Fig. 1.15, and one of them again in Fig. 1.16. These are four leaf eating moths from the coniferous forest at Letzlinger Heide in Germany. Figure 1.15 comes from Varley's (1949) reanalysis

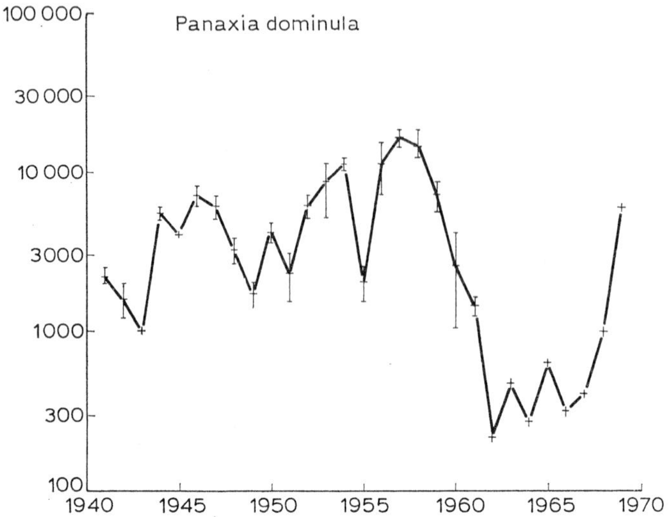

Fig. 1.14 The population of *Panaxia dominula* at Cothill near Oxford. Up to 1961 estimates were made by the capture-recapture method, and probable limits of these estimates are indicated.

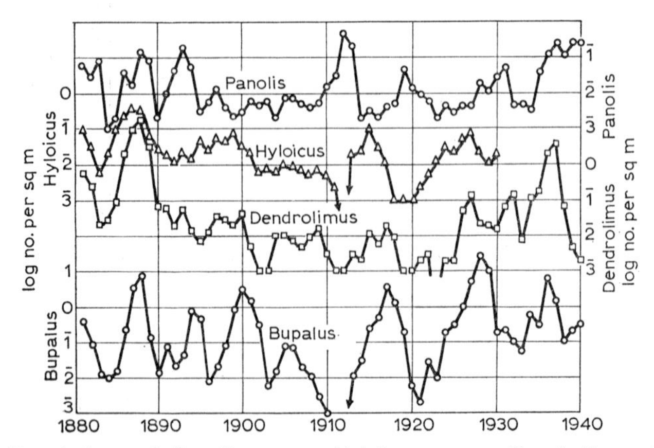

Fig. 1.15 Populations of Panolis pupae, Hyloicus pupae, Dendrolimus hibernating larvae and Bupalus pupae at Letzlinger, in numbers per square metre, plotted logarithmically (from Varley 1949).

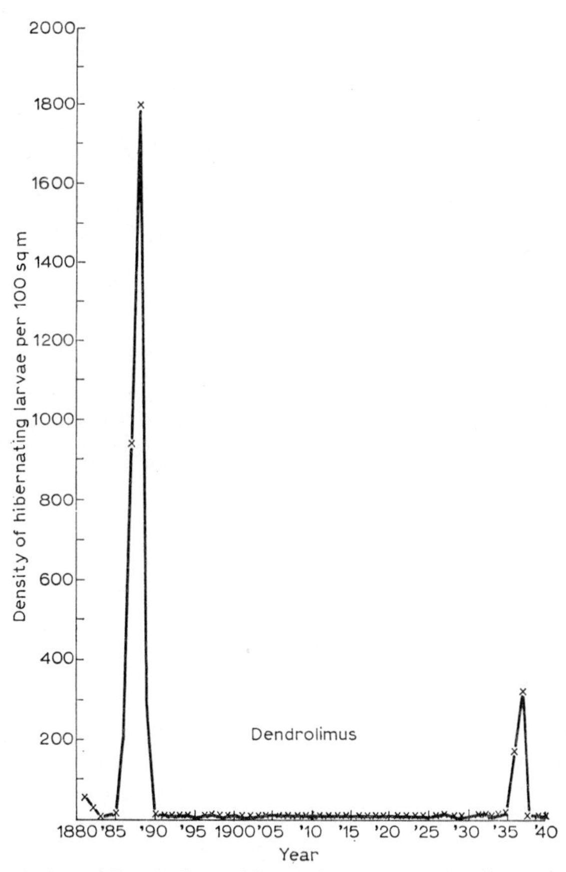

Fig. 1.16 Population of Dendrolimus hibernating larvae at Letzlinger, plotted arithmetically (from Schwerdtfeger, 1941 and Lack 1954).

of Schwerdtfeger's data. These are the longest recorded series for an insect population. Plotted arithmetically the Dendrolimus population appears to have been very constant except for two outbreak periods, and no doubt this is the impression of the population that a casual stroller through the forest might have got. However the logarithmic plot shows that within this apparently steady period the population varied a hundred times; more than the lynx population and appreciably more than any of the bird or plant populations. Table 1 shows that the Bupalus (the pine looper moth) population is the most variable yet recorded, but this may not be typical of the species. Klomp (1966) made a ten year study of this species at Veluwe in the Netherlands, counting all the stages, and these are shown in Fig. 1.17. The line for pupae corresponds to the line in Fig. 1.15 and the standard deviation of this is also given in Table 1.1

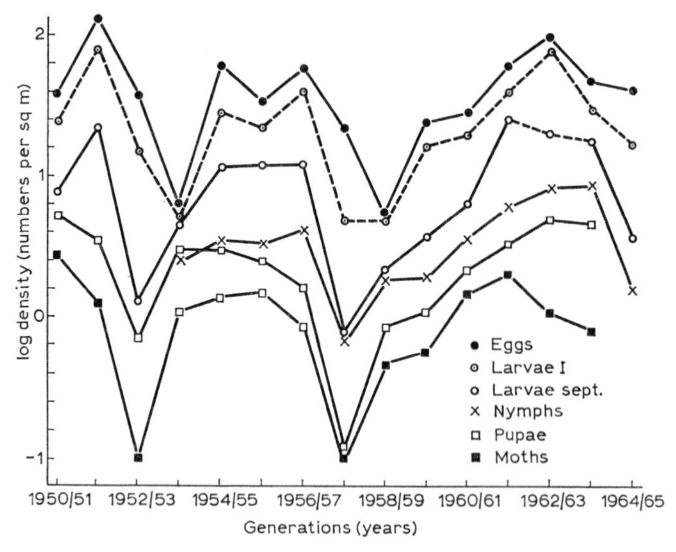

Fig. 1.17 The numbers of various stages of Bupalus per square metre at de Hoge Veluwe (from Klomp 1966).

Complex populations

So far a single number has been used to describe the variation. Most populations can be described more fully by using sets of the numbers for the different classes in populations; for instance the number of males and females, or the numbers of different ages. The next three figures show three populations when some breakdown of this sort have been attempted. Figure 1.18 shows the world population of the gannet (*Sula bassana*). This is a large white sea bird which nests on the cliffs in colonies. All these colonies are very conspicuous and so it is one of the few animals for which it is feasible to count the world population. Some of the lines in Fig. 1.18 refer to individual populations as on the Bass Rock in the Firth of Forth or Ailsa Craig in the Firth of Clyde, but others, as the St. Lawrence line, refer to groups of colonies. The amount of interchange between the colonies is not known, but it is undoubtedly rather small. The changes in the total world population of the gannet seem chiefly to have been caused by overcropping in the 19th century and the recovery from this since. In the 19th century these birds were taken for food and also for oil and feathers. Figure 1.19 shows another population of conspicuous animals, in this case the Pribilof fur seals. These come to the beaches of the Pribilof Islands, in the Bering Sea, for breeding and can be counted then. Here the population can be divided into a number of distinct classes: the harem bulls, the breeding males each with his collection of cows; the idle bulls, the young males which have not yet succeeded in establishing a breeding terri-

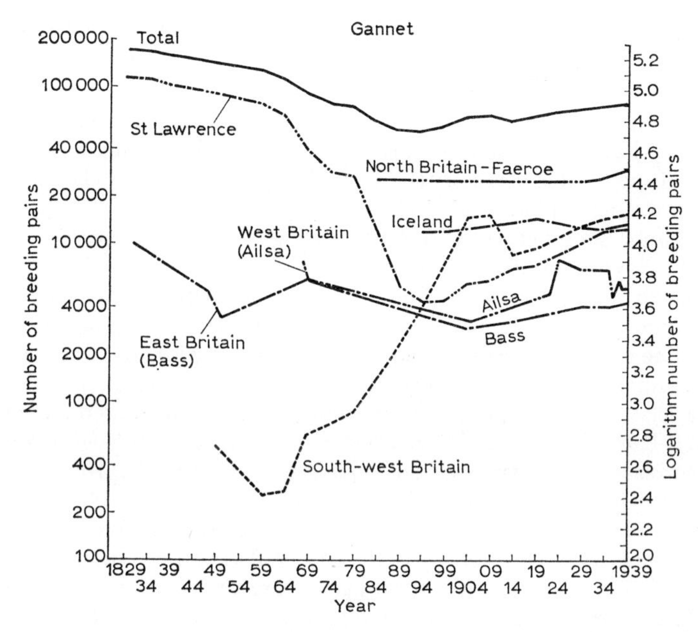

Fig. 1.18 Populations of the gannet, *Sula bassana,* in various areas of the North Atlantic (from Fisher and Lockley 1954, and Fisher and Vevers 1944).

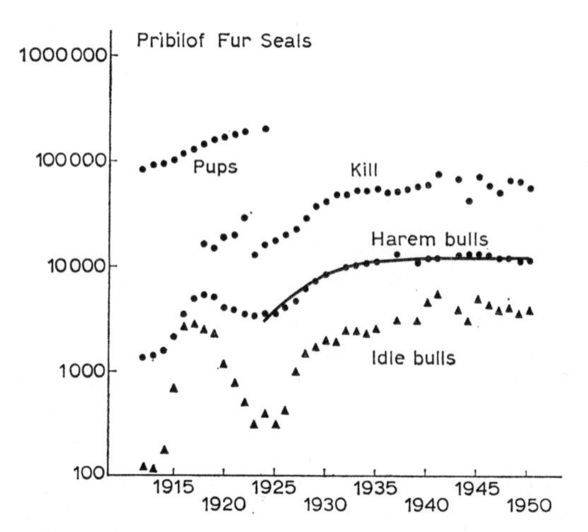

Fig. 1.19 The numbers of various classes of the Pribilof fur seal, *Callorhinus ursinus,* on the Pribilof Islands (from Kenyon, Scheffer and Chapman 1954).

tory; and the pups, the newly born seals of the year. This population was in danger of extinction from over exploitation in the 19th century, but since the beginning of the 20th century has been artificially regulated and maintained, both to produce a valuable yield of skins, and to forestall any danger of extinction. Once again, logarithmic plotting not only shows relative changes in size of the different classes, but is necessary if the numbers of the different classes are to be shown on the same plot.

The numbers of herring (*Clupea harengus*) of different ages are shown in Fig. 1.20. These statistics have been derived from the catches of the fishermen off the north east coast of Scotland, and have been standardized to show the average number of herring caught per boat, per net during the summer season. The top line in the figure shows the total population, while the lines below this show the numbers in the individual year classes. In the right half of the figure, each year class decreases progressively from the time it first entered the fishery. In the left half this is not so. This is surprising as it is not biologically possible for a year class to increase once the herring is four years old. Herring become catchable at either three years old or four years old, so an increase only from three years old to four years old is possible from the biology of the fish in statistics of this sort. The increase in the late 1940's must be ascribed to changes in the population being sampled: the fishermen are known to have been becoming more efficient during this period at finding the concentrations of fish. This was the period in which echo sounders were coming into general use, and it was also a period in which new boats and new nets became available. Looking at the first point in the graph of each year class, it can be seen that there has been something like a tenfold variation in the initial strength of the year classes. On the other hand there has been a much smaller variation, only about twofold, in the total number of herring caught. The variation in the size of the year classes has been compensated by variation in mortality, which is indicated by the slope of the lines of the individual year classes.

A set of species

The final figure in this Chapter, Fig. 1.21 is considerably more complicated than the others. It shows the numbers of 23 different sorts of plankton (Williamson 1961a). These are taken with standard samplers, from the fishing boats catching the herring shown in Fig. 1.20. The numbers are shown on a logarithmic scale along the abscissa. For each species of plankton its average number in each of the 11 years from 1949–59 have been arranged in rank order along the ordinate (or *y* axis). So for each species the highest number is shown at the top of the graph and the smallest number at the bottom and the eleven population estimates are connected by a line. The standard deviations of some of these are given in Table 1.1. It is in any case easy to see, by comparing the position of the top and bottom points of each line, that the variation in numbers for nearly all these species have been between tenfold and a hundredfold. It is also easy to see that the logarithmic

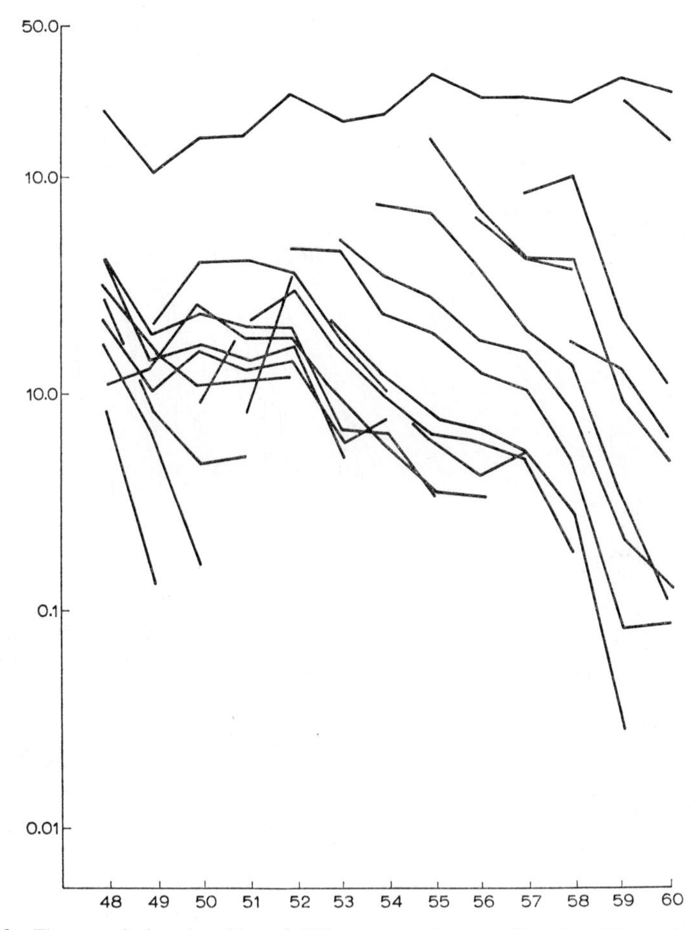

Fig. 1.20 The population densities of different year classes of herring, *Clupea harengus*, caught by Scottish drift-net fishermen in the north-west North Sea, measured in crans, units of volume, per night. The top line is the total catch, the others individual year classes starting with three year old fish (from Williamson 1963).

plot makes the distributions reasonably symmetrical, though it is not so easy to see that it makes them fairly normal. This set of populations will be discussed more fully in Chapter 16.

Figure 1.21 shows not only the variation of the individual species, but the degree of variation between the mean abundance of different species. Most of this has a simple biological explanation. For instance line 15 at the right hand side is a group of phytoplankton, Dinoflagellates of the genus Ceratium. The next group of curves, species 4, 6, 11, 19, 22 and 23 are all herbivorous species, while those further to the left are omnivores, carnivores and species that are abundant at some other time of the year. While a logarithmic plot is desirable for each species individually, it is essential if all the species are to be studied as a set. This is only to be expected when comparing species in different trophic levels, but it makes a fifth reason, in addition to those on p. 4, for using logarithms in population work.

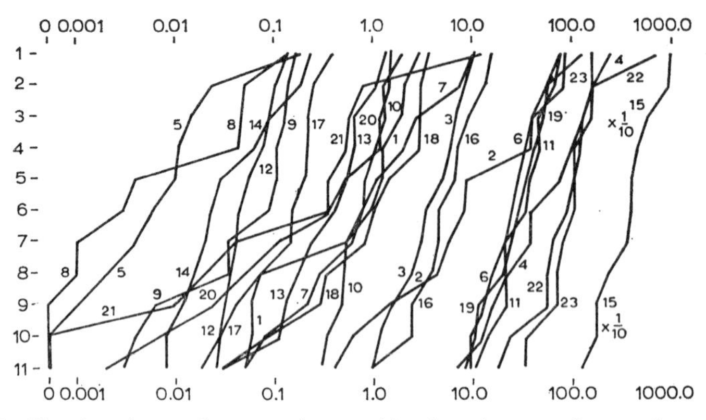

Fig. 1.21 The abundance of twenty-three entities (species, age classes of a species, or species groups) of plankton per third of a cubic metre per summer in the north-west North Sea. The figures for each entity are arranged in rank order, independently of the other entities, from highest to lowest (from Williamson 1961a).

These data show that the definition of the place in which the population occurs is often rather arbitrary, being for instance a wood an ornithologist can observe by himself, or an area of sea in which a set of fishermen work. In only a few cases, such as the moth Panaxia or the gannet populations, can a place be defined less arbitrarily? All populations show some variation, usually rather irregular, and the extent of this varies from a few percent of the mean up to a change of a thousand fold. Rather rarely, populations may show fairly regular cyclical changes as in the lynx. In all cases the logarithm of population size is the most convenient statistic to work with as it shows the relative change in the population, makes the variation symmetrical and almost normal, is frequently necessary to show all the relevant figures on one graph, and as will be shown in the next Chapter is related simply to the basic theory of population dynamics. When several species are studied simultaneously, logarithmic plots are again usually necessary and convenient. Sometimes a single statistic suffices to describe the population size, but many populations are better described by using several figures, and these populations may be called complex. The theory of both simple and complex populations is considered in the next two Chapters.

2 Unlimited population growth: theory and experiments

The problem of analysing the changes in population size of any organism can be split into two parts, that of explaining the mean abundance and that of explaining the variation about this mean. The first Chapter was concerned mostly with descriptions of the second of these. It was noted in passing that the answer to the first, the mean abundance of any particular organism, depended at least to some extent on its position in the food chain, and that the answer to the second, the variation about the mean, can sometimes be ascribed to variations in climate and other physical factors. We will return to the second problem in the next Chapter. This Chapter is concerned with developing deterministic models for both simple and complex populations. These are necessary for understanding how the growth of populations can lead to particular mean abundances, though here we will be concerned only with unlimited growth. This is somewhat artificial, in that the growth of any population must eventually be limited. But the study of unlimited growth is simpler, and is a helpful preliminary to studying limited growth and so leading to a model which allows the interpretation of both the mean abundance and the variation about it.

Unlimited population growth

Consider first a simple organism that divides by binary fission, synchronously. Starting with a single individual this should give:

Time	0	1	2	3	4	5	6
Number	1	2	4	8	16	32	64

From this we can see that n_t, the number at time t is 2^t. By a simple change of variable, $r = ln\ 2$, the number at time t is $(e^r)^t = e^{rt}$.*

If we started with any arbitrary number of individuals rather than one, say n_o, then all the numbers in the table would have been multiplied by n_o, or in other words

$$n_t = n_o . e^{rt}$$

* In this book logarithms to the base *e* will be written *ln*, logarithms to the base 10 log.

If we plot n^t against t we will get a curve of increasing slope which cuts the n axis, when $t=o$, at n_o. A rather more satisfactory plot is to plot $ln\ n$ against t where the curve will intercept the $ln\ n$ axis at $ln\ n_o$ but it will be a straight line with slope r. Growth of this sort is unlimited, exponential and autocatalytic: unlimited because n always increases with increasing t, exponential because of the e^{rt} term, and autocatalytic because the rate of growth depends on the number of individuals already present in the population. This sort of growth will always produce a straight line when plotted on a logarithmic scale of the numbers in the population, and this furnishes the fourth reason mentioned in the first Chapter for using logarithmic plots of numbers against time. Straight sections of this plot show when a population is increasing exponentially and so perhaps without limitation.

This picture of population growth is very simple, but it is frequently satisfactory in practice. Consider now some of the possible complications that would render it invalid. The first is that the population consists in many cases of discrete individuals. That is, n is not a continuous variable but a discontinuous one. In practice the difference this makes is normally extremely small, and can be neglected in relation to other errors that are inevitably to be found in the data, sampling errors and so on. There is one interesting case in the theory of population dynamics where continuity does make a difference, which we will come to in Chapter 14. A second and a much more important complication is that not all individuals are the same. Even with organisms dividing by binary fission, such as bacteria like *Aerobacter aerogenes* or the fission yeast *Schizosaccharomyces pombe*, there will be a variation in the length of each cell between divisions. If the organism grows only in length it will be, say, of unit length just after the division, and will increase in length up to two units just before the next division. So the plot of the length of a single line against time will be a sawtoothed curve going from one to two and back to one again at each division. In practice, cultures do not divide synchronously for very long even when this has been specially arranged, and minor differences of this sort average out. In the first Chapter, in some of the plots of populations there were persistant and important differences among the members of the population. For instance in man there is a non-reproducing stage lasting roughly the first 20 years of life, a reproductive stage from about 20–40 and a post reproductive stage from about 40 to around 100 years old. In greater detail, birth rates and death rates are different at every age and different for the two sexes. These differences can be described in life tables.

Description and theory of life tables

A number of variables can be plotted against the age of the individual, and fuller accounts of these than are needed here can be found in Medawar (1960), Comfort (1964) and in Allee, Emerson, Park, Park and Schmidt (1949). The two that are most usually plotted are the number of individuals surviving starting from a conventional cohort of 1000 against the age of those surviving, and the number of female births for each female in the population. This restriction to females is just for convenience, as it simplifies the mathematics. The first of these, the number surviving to a given age, is usually indicated by l_x, and second by m_x. A plot of log l_x has the advantage of showing the variation in relative death rates. The total reproduction for a cohort is given by

$$R_o = \int_o^\infty l_x m_x dx$$

A simple example will show what is involved. Evans and Smith (1952) studied life tables of the human louse, *Pediculus humanus*, under controlled conditions. The lice were kept on 1½″ squares of woollen cloth and fed daily on volunteers. They were kept between meals in an incubator at 30°C. Figure 2.1 shows the curves for l_x, m_x and for the product $l_x.m_x$ over a 70 day period. It will be seen that the number of females surviving decreases steadily starting from eggs just born, through the nymphal stage between day 9 and 21 until the last adult dies at day 66. Adults start reproducing on day 22. Fertility rises to a maximum around day 40 and then, on the whole, declines slowly while the adults are alive. As $m_x = 0$ on day 21 and $l_x = 0$ on day 67 the product of $l_x.m_x$ must $= 0$ both at day 21 and day 67 and consequently shows at least one peak in between. The total reproduction is the area under the $l_x m_x$ curve, namely $R_o = 30.93$. In this case we have started the l_x curve at 1, and so the meaning of R_o is that for each female egg laid we can expect to get a net production of about 31 eggs. Obviously both l_x and m_x will always depend upon the physical conditions, and typically on the other biological species present. Presumably in this case the l_x curve is determined mostly by accidents to individuals, though if it is determined entirely this way one would expect a negative exponential curve with rather greater curvature than that shown in the figure.

The mathematics of the life table is normally handled by the differential calculus. As we have seen this is an approximation. The lice, in this case, can only be counted at a particular time, and there must be a discrete number of them. In practice these discontinuities are small and can be ignored. However when we come to the practical calculation of figures from life tables we find that demographers in fact work with discrete numbers. For instance the demography of man is usually done using five year intervals. Consequently the true representation both of the population and of the calculations as they are usually done requires discrete mathematics rather than continuous mathematics, that is to say the mathematics of matrices. Those unfamiliar with matrices will find a summary of their properties in the Appendix.

Leslie Matrices

The use of matrix notation to show the growth of complex populations has largely been developed as the result of the work by Leslie (1945, 1948). This

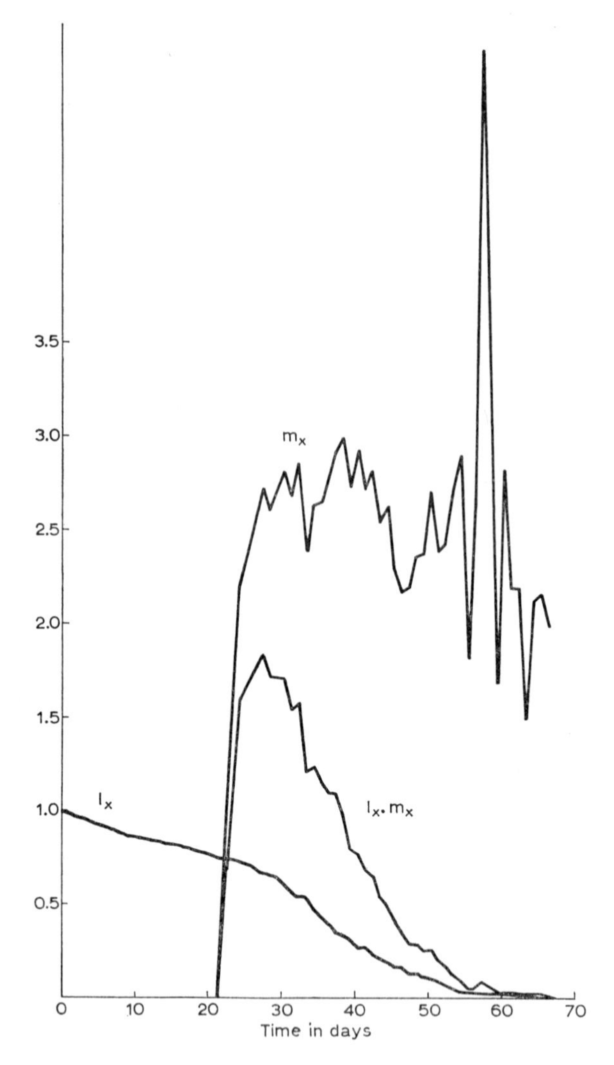

Fig. 2.1 Life table statistics for the human louse, *Pediculus humanus* (from data in Evans and Smith 1952).

notation was also invented independently, but not developed so far, by Lewis (1942). Descriptions and extensions of Leslie's work with matrices will be found in Moran (1962), Pielou (1969), Usher (1966), Usher and Williamson (1970) and Williamson (1959, 1967) and the papers referred to by these authors. I shall return to Leslie matrices from time to time in later Chapters as they represent one of the most convenient forms of writing down population systems for practical calculations.

The basic Leslie system for describing complex populations sets out the numbers of the individuals in the different classes of the population as a column vector, and premultiplies this by a transition matrix.

$$\begin{bmatrix} f_o & f_1 & f_2 & \cdots f_{k1} & f_k \\ p_o & 0 & 0 & \cdots 0 & 0 \\ 0 & p_1 & 0 & \cdots 0 & 0 \\ \cdots\cdots & & & \cdots & \cdots\cdots \\ 0 & 0 & 0 & \cdots p_{k-1} & 0 \end{bmatrix} \begin{bmatrix} n_o \\ n_1 \\ n_2 \\ \cdots \\ n_k \end{bmatrix}$$

The numbers in each class of the population are one unit older each row further down the column vector, and this unit is whatever unit is convenient for the particular population being studied. So n_i is the number in the ith age class, p_i is the survival factor from the ith to the $(i+1)^{th}$ class and f_i is the mean fertility of the ith class, i.e. the number of newly born individuals expected to be produced during one time interval for each member of the ith class at the beginning of the interval. It will be seen that the f's run across the top line of the matrix because they give the transition to the 0th class. The p's run down the sub-diagonal one step below the principal diagonal, because there is a transition one step down the column vector at each stage. All the other elements of the matrix are zero, as they refer to biologically impossible transitions. The p's being probabilities can vary between zero and one, while the f's being fertilities are greater than or equal to zero, but have no particular upper bound. So all elements are either zero or positive. The transition matrix is a $(k+1)\times(k+1)$ square matrix and the column vector is $(k+1) \times 1$ matrix. The transition matrix necessarily has $(k+1)$ latent roots. The sum of these is equal to the trace of the matrix, the sum of the elements of the principal diagonal. As f_o, the fertility of the youngest class will normally be zero, the sum of the roots will also normally be zero. From the Frobenius-Perron theorem,* as the elements of the matrix are non-negative, the dominant root will be real and positive. Provided that there is at least one non-zero f in addition to f_k, no other root will be of the same modulus as the dominant, so the population will eventually settle down to a distribution, the stable-age distribution, given by the dominant latent vector, and the rate of increase of the population will be given by the dominant latent root, which we may call λ_1. This can be shown by an example.

* This and some useful theorems derived from it are mentioned in Usher (1966) and Usher and Williamson (1970).

If we take the system

$$\begin{bmatrix} 0 & 9 & 12 \\ \frac{1}{3} & 0 & 0 \\ 0 & \frac{1}{2} & 0 \end{bmatrix} \begin{bmatrix} 24 \\ 4 \\ 1 \end{bmatrix} \rightarrow \begin{bmatrix} 48 \\ 8 \\ 2 \end{bmatrix}$$

we have a system with three age classes which can be labelled young, middle-aged and old. The net fertility of the young is 0, for the middle-aged is 9, and of the old is 12. The probability of survival from young to middle-aged is one third and from middle-aged to old is a half. Each element on the right hand side is exactly twice the corresponding element in the vector of the left hand side. That is to say the number in each age class, and in the population as a whole, has doubled in the time of one transition. As the proportions between the three classes are the same in both vectors, both vectors show the stable age distribution, which is defined in relative terms and not in absolute ones. The dominant latent root is given by the ratio of the elements to the second vector to the first and therefore $\lambda_1=2$. Under this transition matrix, the population in its stable distribution doubles in each time interval.

There is a simple relationship between the λ of the Leslie matrix and the r that gave the rate of increase for a simple population. Write N_i for Σn_i, the total complex population, and λ for λ_1, then

$$N_{j+m}=\lambda^m N_i,$$

provided the population was in the stable age distribution at time j. With $j=0$ $m=t$ this becomes

$$N_t=\lambda^t N_o,$$

and this can be compared with

$$n_t=n_o e^{rt}$$

of the simple case.
From this $\lambda^t=e^{rt}$ or $\lambda=e^r$, $r=ln\ \lambda$.
Neither r nor λ are measured in absolute units; both are defined in relation to a particular unit of time. We can define $r\equiv ln\lambda$ for the complex case and so have a parameter which corresponds exactly to that used in the simple case. In both cases, r, or λ, will depend on the physical conditions.

There is one important difference in the complex case: the rate of increase observed will depend not only on the physical conditions, and so on the r that would eventually be reached, but also on the age distribution of the population. Using the same transition matrix if we had a column vector with just one adult namely $[0\ 0\ 1]^T$ then at the next time interval the vector is $[12\ 0\ 0]^T$ or an increase of 12 fold in the total population size. On the other hand from $[12\ 0\ 0]^T$ the vector goes to $[0\ 4\ 0]^T$, a decrease to 1/3 in the total population size. These

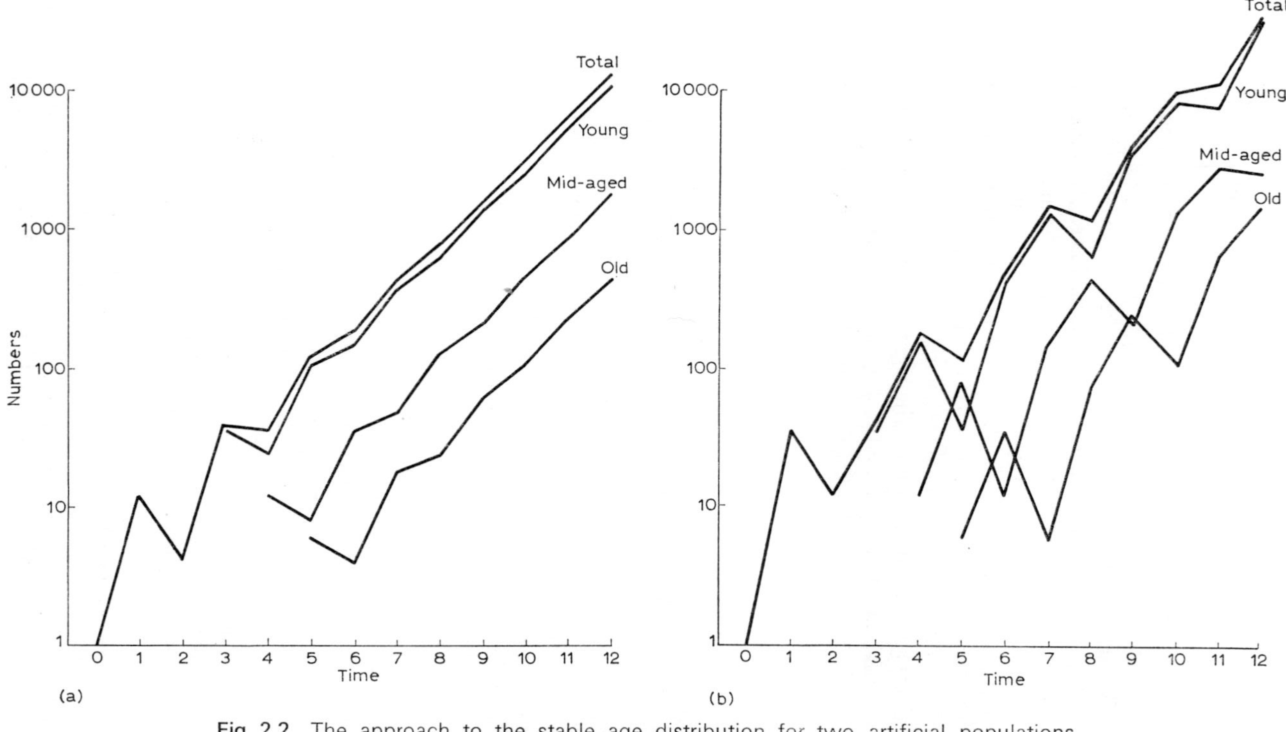

Fig. 2.2 The approach to the stable age distribution for two artificial populations generated by Leslie matrices (from Williamson 1967).

are the extreme possibilities for this particular transition matrix, but they show that the population will only increase at the rate r when it is in the stable age distribution. In all other distributions the rate of increase may be either greater or less than r.

As the Leslie matrix has $k+1$ roots, which normally sum to 0 and as the dominant root is real and positive, the other roots will be either negative or complex. It can be shown that no other real positive root is possible, though some of the complex roots may have positive real parts. These other roots do not come into the stable age distribution or the rate of increase in the stable age distribution, because these are determined entirely by the dominant root and dominant vector. What the other roots determine is the rate of approach to the stable age distribution. This is shown in Fig. 2.2. The left hand half shows the rate of approach to the stable age distribution starting with one breeding individual for the transition matrix given above. The figure on the right shows the rate of approach to the equilibrium for the transition matrix

$$\begin{bmatrix} 0 & 3 & 36 \\ \frac{1}{3} & 0 & 0 \\ 0 & \frac{1}{2} & 0 \end{bmatrix}$$

which has the same dominant latent root, $\lambda=2$, and the same dominant vector $[24\ 4\ 1]^T$. These matrices differ only in the second and third roots. As the sum of the roots necessarily adds to zero only the second need be considered. For the first transition matrix it is -1, for the second $-1+1.414i$, with a modulus of 1.732. The system is quite clearly much less damped in the second case, and this is a reflection of the fact that the second latent root is nearer in modulus to the dominant latent root in that matrix.

The components of the population listed in the column vector can be any components of the population and need not be age classes as in the original Leslie matrix. They can be size classes of trees (Usher 1966) or different genotypes in a population (Williamson 1959) or any other useful subdivision of the population. The theory given above leading to a stable distribution of classes and a stable rate of increase will apply so long as the different classes do not interact with each other. In the genetic case this is not true. Heterozygotes can be formed either by matings between heterozygotes or by mating of one homozygote with a different homozygote. Consequently as Feller (1967) pointed out, using a rather different derivation, Mendelian populations will not have a stable rate of increase. In practice though, the deviations from the stable rate of increase will be very small, and then can be neglected. Genetic matrices are considered in Chapter 6 (p. 65).

The usefulness of r or λ

The parameter r, which is equal to $ln\ \lambda_1$, has been given a variety of names. Lotka (1925) called it the 'natural' rate of increase and later the true and the inherent rate of increase. Andrewartha and Birch (1954) who use the symbolism r_m call it the innate capacity for increase; Fisher (1931) the malthusian parameter; but Lotka's (1945) later term, the intrinsic rate of natural increase, is that generally used by ecologists, though geneticists frequently use Fisher's term. There are four points that may be noted:

1 It refers to a particular time interval, it is not a pure number.

2 It refers to a particular set of birth and death rates, not necessarily an optimal set.

3 It includes both births and deaths.

4 There is no feedback from the population size included, it it a density free parameter.

It is merely the maximum steady rate of increase under particular conditions, before the population density starts affecting birth and death rates. It is undoubtedly a useful statistic in some circumstances, for instance it has been used for comparing genotypes by Dobzhansky, Lewontin and Pavlovsky (1964), but of points 2 and 3 make it of somewhat limited use. Births and deaths are quite distinct biological processes, and a parameter that includes both of them cannot hope to be as useful as separate estimates of the variation in birth and death rates. A knowledge of the intrinsic rates of increase does not for example tell one how the population will behave when it is exploited (Williamson 1967). Its advantage is largely that it is a single number, and it is usually easier to think about one parameter at a time than many. To give an indication of the numerical size, and a comparison of r and λ, the following table compares the rate of increase of the beetle *Tribolium castaneum* and the rat *Rattus norvegicus:*

Species	r (day)	r (week)	λ (week)	Time to double (days)
T. castaneum	0.101	0.707	2.03	6.86
R. norvegicus	0.0147	0.1029	1.11	47.14

This table shows one advantage of r over λ. The figure for a week is got from that for a day simply by multiplying by 7. On the other hand the last two columns show that the information given by λ is more easily related to one's everyday experience. As *T. castaneum* increases by slightly more than 2 times within a week it not surprisingly takes slightly less than a week to double, and correspondingly, *R. norvegicus* which increases 11% in a week takes slightly less than 7 weeks to double its population size. The 'time to double' is the positive equivalent of the 'half-life' used by Harper (1967) for some cases of exponential decrease in plant populations.

Both r and λ depend on physical conditions. The only variables in the population's environment that they eliminate are the effects of the population's own density and the population's own age structure. This is shown in Fig. 2.3 from

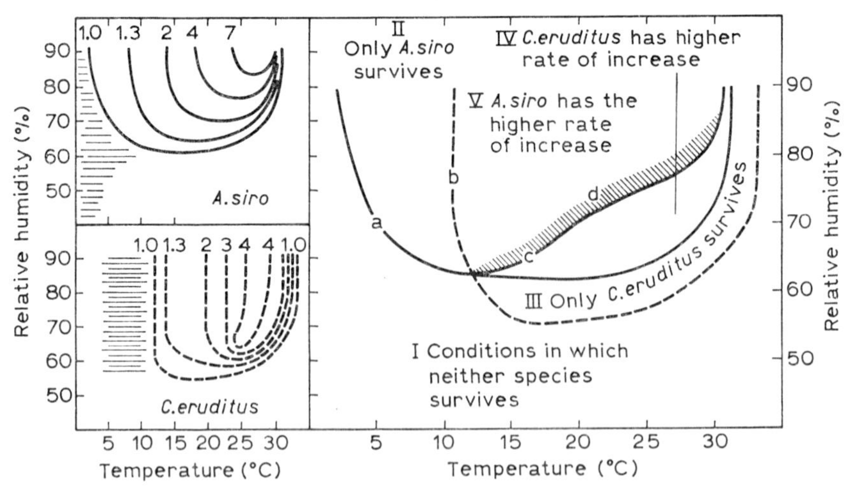

Fig. 2.3 Variation of the multiplicative rate of two species of mites with temperature and humidity (from Solomon 1962).

Solomon (1962). This figure shows λ per week for two mites living in flour under different conditions of temperature and humidity. *Acarus siro* feeds on the flour, *Cheyletus eruditis* feeds on *Acarus siro*. The two left hand diagrams show that the species have rather different optima. Acarus likes a higher temperature and can increase by over 7 times per week. Cheyletus prefers a lower temperature and is less sensitive to variations of humidity but can only increase by 4 times a week. The right hand half of the diagram shows that there is a moderately limited range of conditions in which Cheyletus has a higher rate of increase. The effect of Cheyletus preying on Acarus has not been allowed for in calculating the rate of increase in Acarus, so there will be a zone indicated by the hatching marked D in which Cheyletus will continue to be an effective control on Acarus even though Acarus has a higher λ. These facts bring out two important points. The first is that the rate of increase is dependent on the physical conditions. The second is due to Smith (1961) who pointed out that under unlimited growth there will be a relationship between the rate of increase and the physical conditions. Conversely, as will be shown in the next Chapter, when there is a relation between population size and the physical conditions, there is density dependent population control.

3 Limited population growth: theory and experiments

In the last Chapter we were concerned with unlimited population growth which leads to an exponential change in population size. This may be either positive or negative. In the example used there the changes were positive, and if the populations had continued to increase in the way shown they would have gone to an infinite population size. This is clearly impossible, and applies however slowly the positive increase may be. Darwin (1859) pointed out that a pair of elephants, a rather slow breeding species, would nevertheless leave fifteen million live descendents after only five hundred years. So there is no doubt that population growth must be limited when r is positive, which is the same as saying when λ is greater than 1. r can be negative, or in other words λ can be less than 1, as can be seen from Solomon's example in Fig. 2.3. In such a case the population will decline to extinction unless there is a change of this rate of decline. At one time there was a fierce dispute as to whether the limitation of population growth must necessarily be related to the population density. That this is so, and that this necessity is quite independent of any random fluctuations there may be, has been shown neatly both by Haldane (1953), and Moran (1962). The contrary view was urged by Andrewartha and Birch (1954) though with some interesting inconsistencies on page 32, but since Birch's paper (1962) the argument can be considered settled. Birch now accepts that populations will frequently be limited in a density dependent way, though he argues there will be some exceptions. This is a perfectly acceptable point of view, because the demonstration by Moran and Haldane refers only to populations that persist. There is no reason why some populations should not be temporary ones, and as such not subject density dependent limitations. So this Chapter is concerned with showing precisely what is meant by density dependent limitation and how it works when it does work, rather than arguing about whether or not it works.

When there is density dependent limitation there will be a relationship between the rate of change of the population and the population size. If we define the change between two populations censuses n_{i+1} and n_i as Δn_i then there will be a relationship between Δn and n. As with the relation between λ and physical conditions in Solomon's mites, this will not necessarily be a linear relationship but it will in this case normally be a monotonic one. That is to say the rate of change of population size will change consistently, with the same sign, with the change of population size. An example of this is shown in Fig. 3.1, which is some data on oven birds given by MacArthur and Connell (1966). It can be seen that the population tends to decrease when high and increase when low. This pattern is normal in many populations, for instance it has been shown in plants, and

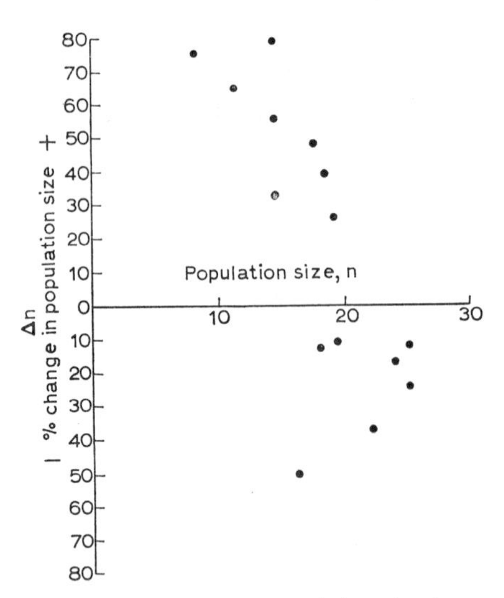

Fig. 3.1 Change of population size against population size for an ovenbird population near Cleveland, Ohio, showing an equilibrium at a population of about twenty (from MacArthur and Connell 1966).

Harper (1967) compares data for *Plantago*, *Anthroxanthum*, *Matricaria* and *Papaver*. Another way of expressing the same sort of change, is that the number of new individuals born into the population will tend to bring the population back to the same size. In extreme cases this will mean that the number of individuals born into the population will be the same irrespective of population size, and a case in point is given by Beverton (1962) for the plaice in the southern North Sea.

There will of course be populations which are temporary and do not persist indefinitely. For instance Reynoldson (1948) showed that the small oligochaete *Enchytraeus albidus* was washed out of the Huddersfield sewage beds in summer after heavy rain. Even in this case, there is some indication of density dependent effects at high population densities between heavy rain storms. As a rule of thumb, we can say that any population which persists for ten generations or more is likely to show density dependent effects. The fact that the population may become extinct in 100 or 1000 or more generations does not affect the validity of the density dependent control during the period in which it is maintaining itself. So we need not consider whether or not there is density dependent regulation of populations but merely how to define density actions and to show the way in which they would interact.

Basic theory of population dynamics

There are only four factors which can change the size of a population. These are births, deaths, immigrants and emigrants. For those populations in which the place in which they occur is arbitrarily defined, immigrants and emigrants must obviously be considered. In those where the place has a natural definition, immigrants and emigrants may still be important. There may be a few populations for which one or more of these terms will always be zero, and some of these are discussed in the next Chapter, but no population can have more factors than these. Factors can be combined to form a fundamental equation

$$N_{i+1}=N_i+B-D+I-E$$

None of the terms in this equation is constant: all of them vary depending on conditions. Subtracting N_i from both sides we get

$$\Delta N_i=B-D+I-E,$$

and dividing through now by N_i to make them relative rates of change,

$$\Delta N_i/N_i=b-d+i-e.$$

Over a reasonable period of time, say ten generations or more, the left hand side will have to approach zero, and for this to happen, because of the constitution of the right hand side, some or all of b, d, i, e, must be functions of N. This is the essential point in Haldane's and Moran's argument. The nature of these functions is considered below.

The same equation can be written in continuous form, by noticing that as the interval of time decreases ΔN approximates to dn/dt and so

$$\frac{dn/dt}{n}=\frac{d\ln n}{dt}=b-d+i-e.$$

This equation, which is another form of the fundamental equation of population dynamics, gives yet another reason for using the logarithm of the population size as the primary variable when studying populations.

The theory is most easily discussed if we take $i=e=0$ to start with and just consider b and d. The way i and e affect the population density when they are not zero follows naturally once the simpler theory has been developed. Some of the possible relationships of b and d to density are shown in Fig. 3.2. b and d are plotted against the ordinate (the y axis) while population density is plotted along the abscissa (the x axis). Diagram A shows both births and deaths running at the same value irrespective of the population density, and this defines what is meant by density independent. In diagram B the rate increases with population density. The increase is not necessarily linear, but must be monotonic in the range of density in which we are considering it. Such a change with density will tend to regulate the population if it applies to the death rate, but tend to make it unstable if it refers to the birth rate, as will be shown below. Diagram C shows the reverse situation. In diagram D the birth rate is density independent but the death rate is density dependent, and for simplicity is shown as a simple linear change, but any monotonic change would suffice. At the point where the two curves intersect,

$b=d$, and consequently from the fundamental equation, $d \ln n/dt=0$, that is the population is in equilibrium. This is shown in diagram D by the point $\hat{n}$ on the x axis. Other simple cases of stable population regulation are shown in *E* and *F*. In *E* there is a density independent death rate and a density dependent, that is to say decreasing, birth rate, again leading to equilibrium at $\hat{n}$. In *F* there is a monotonically increasing death rate and monotonically decreasing birth rate which leads again to equilibrium at $\hat{n}$. In all three cases, given by diagrams *D*, *E* and *F*, for populations to the left of $\hat{n}$, where the population is less than the equilibrium population density, b is greater than d. Consequently the populations will tend to increase towards $\hat{n}$. Similarly to the right of $\hat{n}$ the population size is above the equilibrium population density, death rate is

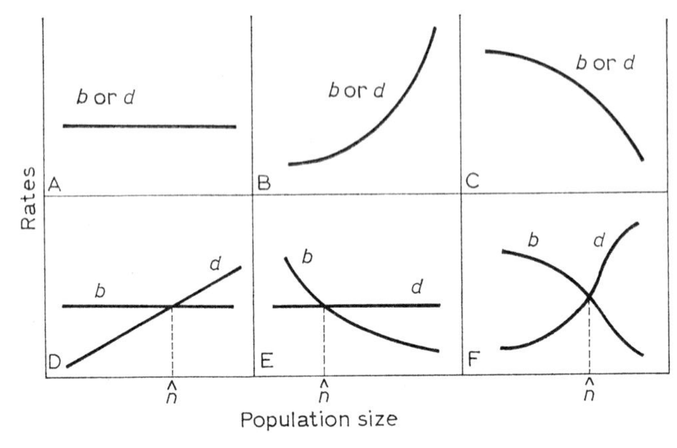

Fig. 3.2 Variation in rates of either births or deaths at different population sizes. $\hat{n}$ is an equilibrium population size. For further explanation see the text.

higher than the birth rate, and so the population will tend to decrease down to the population equilibrium point. The population does not necessarily approach the equilibrium point smoothly, it may oscillate about it, and this will be considered further in Chapter 5 with the aid of Moran diagrams. Nevertheless these diagrams show the essentials of density dependent population control. In the next Chapter it will be shown that the situation of Fig. 3.2 D corresponds to the continuous culture system known as turbidostat, while Fig. 3.2 E corresponds to a chemostat. Before considering complications in these diagrams, it will be useful to discuss in detail one particular important type of limited population growth.

Sigmoid population growth

Many populations when introduced into a new locality show what is known as sigmoid population growth. That is to say the population numbers when plotted arithmetically against time show an S-shaped curve, which shows that this is a slow rate of increase to start with (dn/dt is small), leading to a faster rate of increase at rather greater population size (dn/dt larger), flattening off to an asymptote eventually ($dn/dt=0$). This phenomenon is illustrated repeatedly in standard text books of ecology and so it is only shown diagramatically here in Fig. 3.3. The right hand half of Fig. 3.3 shows the same curve when one plots

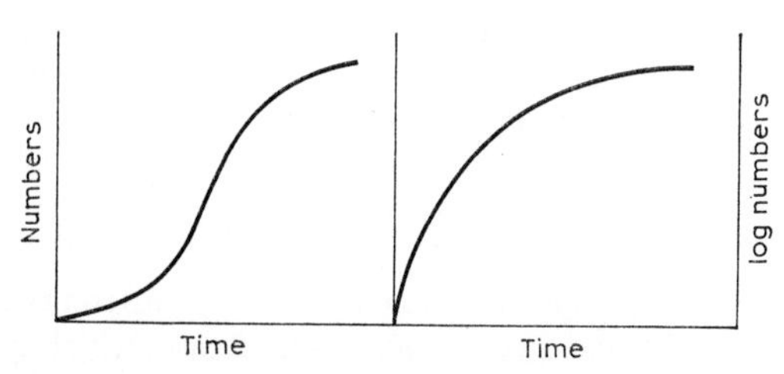

Fig. 3.3 The logistic population curve. The theoretical curve for numbers against time on the left, and for the logarithm of numbers against time on the right.

the logarithm of population size against time, In this form the curve has no inflection, with the early growth approximating to a straight line. This is because the region of the lower inflection of the arithmetic plot corresponds roughly speaking, to exponential growth ($d\ ln\ n/dt$ constant). In either case, three parameters are needed to describe this curve. The easiest way to define these parameters is to have:

1 for the slope of the initial part of the logarithmic curve,
2 for turning the curve over to reach its asymptote, and
3 for fixing the position of the curve in relation to the t axis.

The three parameters can be defined in a number of other ways, provided that together they define a curve of this sort, but whichever way is chosen, three will always be needed. The most popular mathematical expression for fitting to these curves is the equation known as the logistic equation. Feller (1940) showed that the logistic was by no means necessarily the best fit to some of the published data. He showed that two other expressions, also using three parameters, were at least as good in some cases and possibly better in others. For complex populations it is usually easy to show that the logistic equation does not really get a satisfactory fit. For instance Sang (1950) showed this in great detail for Drosophila. Nevertheless the logistic is a simple empirical formula to fit curves of this sort, and it is used in more elaborate forms in later parts of the theory, so it is necessary to discuss it here. It has been used in some interesting discussions of genetic effects in populations by Levins (1968) and MacArthur and Wilson (1967). Whenever it is used, it is necessary to remember that it is a rather crude first approximation, and consequently any conclusions which depend on it being exactly true are very unlikely to be correct. Fortunately the studies using logistic curves and their derivatives are usually not dependent on the exact form of the equation.

The logistic equation is a simple elaborationof one form of exponential growth. In Chapter 2 exponential growth was written $n_t = n_0 e^{rt}$. Differentiating this gives $d \ln n / dt = r$. The logistic equation is the simplest density dependent elaboration on this and is

$$\frac{d \ln n}{dt} = r - cn,$$

where r is, as before, the intrinsic rate of natural increase, and c is another constant. The equation is not usually written in that form, even though it is the simplest. The usual form is derived as follows

$$dn/dt = n(r - cn) = rn - cn^2 = rn\,(1 - cn) = rn\left\{\frac{K - n}{K}\right\}$$

This introduces a parameter K which is $1/c$. In this differential form there are two parameters, r (the intrinsic rate of natural increase) which gives the slope of the initial part of the curve, and K which will be found to be equal to the equilibrium value of n. It is important to note that K is an asymptote, a balance point, and not a limit to the population. That is to say if n is artificially increased above K, and there is usually no reason at all why this should not be done, the population will decline towards K. The third parameter comes in when the logistic is written in integral form:

$$n_t = \frac{K}{1 + \exp\,(z - rt)}$$

where z is a constant involving n_0. Again, in the usual convention, both r and c (or K) include both births and deaths. However if we allow that r, which is a positive term, relates only to births and c, which is a negative term, relates only to deaths then the equation represents the situation in Fig. 3.2 D.

The relation of density dependent theory to changes in physical factors

All parameters of population growth, whether they are basic ones like birth and death rates, or derived ones such as r and K of the logistic equation, will normally vary depending on the physical conditions in which the population finds itself. This variation in physical conditions is summed by Andrewartha and Birch under the heading 'weather', which is no doubt appropriate for terrestrial animals. It is rather less appropriate for terrestrial plants and not at all appropriate for aquatic organisms or most microbes. Irrespective of what these physical factors are called, the effects of their changing on rates of change on population size and equilibrium population density have caused considerable confusion. Klomp (1962) has been into the matter in some detail, but the basic considerations are quite simple. They are easily derived from the diagrams in Fig. 3.2. To take a simple case, suppose that the death rate is density dependent and does not vary with conditions, while the birth rate is density independent but does vary with conditions. This gives the situation in Fig. 3.4 where d shows the death rate with a fixed relationship to population density and b_1, b_2, b_3 the birth rates under three different physical conditions. It will be readily seen that there are three equilibria at population densities $\hat{n}_1$, $\hat{n}_2$, $\hat{n}_3$ corresponding to the three different birth rates. That is, there will be a relationship between the equilibrium population size and the conditions. Most other possible changes in the population can be considered by similar elaborations of Fig. 3.2. For instance if an extra death rate is added,

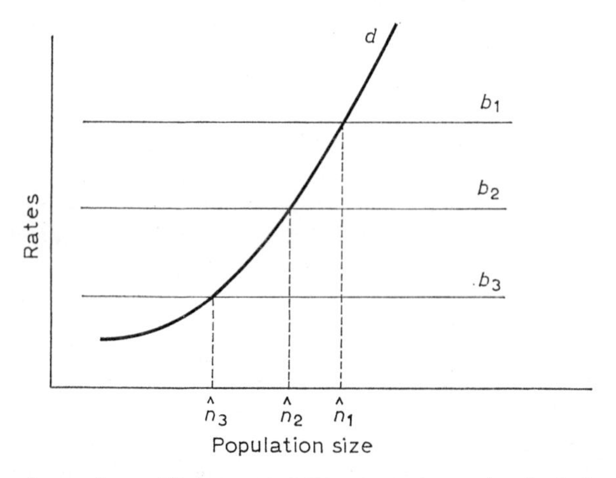

Fig. 3.4 The change in equilibrium population size, when a density independent birth rate varies, and a density dependent death rate does not. d is the death rate, b_1, b_2, b_3 the three birth rates, $\hat{n}_1$, $\hat{n}_2$, $\hat{n}_3$ the corresponding equilibrium populations.

this will tend to make the total curve for d go higher up the diagram, and consequently move the equilibrium population size to the left. Similarly immigration will tend in general to increase the equilibrium population size, while emigration will tend to decrease it. Manipulation of these diagrams also shows the conditions under which these general rules may not hold. For instance, if one of the rates is described by a function which has a large step in it instead of increasing smoothly as in all the diagrams shown here, then the equilibrium population size may not vary when some of the other rates are varied (Williamson 1958). This will however be extremely unusual. In general, any change in birth rate, death rate, immigration or emigration rate will produce a corresponding change in the equilibrium population size. The equilibrium population size will be a function of these four rates and, as these rates are themselves functions of the physical conditions, will be a function of the physical conditions. At the end of the last Chapter Solomon's data exemplified Smith's first proposition, that when there is unlimited growth the rate of change of population size is related to physical conditions. Smith's second proposition is that when the equilibrium population size is related to the physical conditions then there is a limited population growth.

The nature of density dependent factors

While it is not particularly difficult, and in view of the simplicity of the theory scarcely necessary, to show that there are density dependent effects on populations, it is very much harder to show what factors are causing the density dependent effects. Curves of Δn against n such as are shown in Fig. 3.1 show the existence of density dependence, but give no indication of where it comes from. Similarly, Harper (1967) has said 'this self-regulating property of plant population is becoming increasingly well documented although the ultimate causes of density-induced deaths are still very obscure'.

Possible density dependent factors can be classified into, firstly, enemies (whether predators, parasites, disease, or other attacks of one organism by another), secondly food and other similar requirements of the organism, and thirdly, self limiting systems, of which the territorial behaviour in birds is the best known. Which of these are important, and it is quite possible for more than one to be important for any particular population, can only be shown by detailed measurements of the factor and the change of population size. So far it has been difficult to produce conclusive evidence for one factor, particularly when there is only observational data available. Beaver (1967) showed that predators were important density dependent regulators of the bark-beetle Scolytus at low densities, while parasites appeared to be more important at high densities. Harper (1967) quotes work by his team that showed that seedling mortality was important, and this was presumably caused by herbivores. In sock-eyed salmon, Johnson (1965) showed that juvenile mortality from enemies was density dependent. Lack (1966) has argued that food is frequently the most important density dependent factor for birds, and in a somewhat more quantitative and experimental way Reynoldson (1966) has shown that it is an important density dependent factor in lake-dwelling planarians. Southern (1970) has shown that territory has an effect on tawny owl populations, though this of course does not necessarily contradict Lack's views, as both territory and food could be important simultaneously, while Pontin (1961) has shown the effect of territory in ants. Pontin's work will be discussed more fully in Chapter 12. Broadhead and Wapshere (1966) show there is apparently intraspecific interactions between the psocids *Mesopsocus* for egg laying sites on larch twigs, while Klomp (1966) suggests that for *Bupalus* at Veluwe interactions between caterpillars on twigs are important. Johnson suggests density-dependent interactions between sock-eye salmon adults. Nevertheless these studies do not usually lead to equations showing the functional relations of factors, to models simulating the population system. Broadhead and Wapshere produce a model for the main factors in *Mesopsocus* populations, and Pennycuick (1969) analyses the work from Lack's team on the great tit in Marley Wood to show some mathematical relationships there. Both these studies show the importance of density dependent factors in models using parameters derived from observations of real populations, and are interesting on that account. Some details of Pennycuick's model are criticised by Krebs (1970). These can only at this stage be first steps in building full models which show the quantitative relationships of various factors to population density, and the relation of population size to physical factors.

Simple analysis of changes in single populations

Considering the work involved in getting reliable statistics of population size, it is perhaps not surprising that the analysis of change in population size has not progressed further. Some more elaborate methods of analysing observational data will be considered in Chapter 5. This Chapter is concerned primarily with the theory and experiments on basic population growth, and this will be completed by considering a famous laboratory experiment on a simple organism under simple conditions. G. F. Gause did a number of experiments with laboratory populations of microorganisms during the 1930's. Most of these are reported in Gause (1934, 1935) and some important ones have been re-analysed by Leslie (1957). Some of these experiments will be considered in Chapters 10 and 14. Here we are concerned only with a simple preliminary experiment on population growth, to which a number of authors, including Gause and Leslie, have fitted the logistic curve.

Paramecium aurelia is a ciliate protozoan which grows normally by binary fission. Gause grew it in a simple mixture of salts known as Osterhaut's medium, in five cubic centimetres of fluid at 26°C, and did three replicates of his population starting each with 20 individuals. The populations were fed on the pathogenic bacterium known now as *Pseudomonas aeruginosa*, but in Gause's time as *Baccillus pyocyaneus*, which is a bacterium that does not grow in Osterhauts medium. The Paramecia were separated from the detritus and re-suspended in fresh medium each day. The logarithm of the population size against time for the three replicates is shown in Fig. 3.5, and also a plot of Δn against n. At first sight, the curve suggests that the population fits a logistic equation quite neatly. Leslie has recalculated the parameters of the logistic equation allowing for the particular factors of the experiment, daily growth and sampling. The consequences of his calculation are shown in the lower graph where the continuous line shows the expected relationship between Δn and n. The population should come to equilibrium where the curve crosses the $\Delta n=0$ axis. It can be seen that this corresponds quite well to where Δn changes from positive to negative. However, just to the left of this point most of the observed observations are well above Leslie's line, while just to the right they are mostly well below it. Leslie's curve is in fact rather a bad fit to the points, and it is not an unreasonable presumption to suggest that the logistic curve is in fact inappropriate to these data.

There is one important feature of Gause's experiments, which is mentioned by Gause (1934) on page 109, but which has usually been overlooked. Leslie mentioned it in deriving his equations, but paid no particular attention to it.

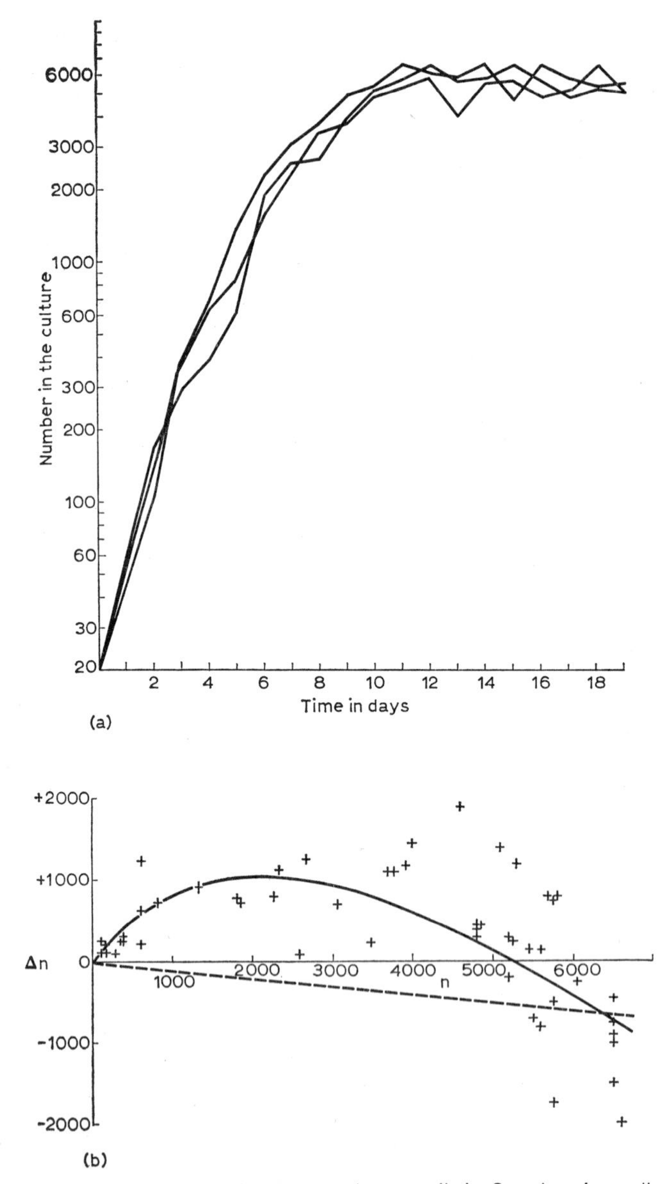

Fig. 3.5 Gause's experiments with *Paramecium aurelia* in Osterhaut's medium.
(a) the logarithm of the population size, in each of three replicates, against time.
(b) the change of population size, Δn, against population size, n, for all three replicates. The solid line is the theoretical logistic curve from Leslie's (1957) calculations. The dotted line is a 10% decrease.

This is the fact that Gause removed one tenth of his population each day in order to count it and did not return it. Consequently, if the Paramecium population did not increase, Gause's handling of it would make it decrease by 10%. It can be seen in the right hand half of Fig. 3.5 that the Δn figures are somewhat near 10% of n. The simplest interpretation of Fig. 3.5 is in fact that the Paramecia are still increasing up to the apparent equilibrium size, but then on many days have suddenly ceased growth. The result of this is to give a steep curve of Δn against n near the equilibrium point, resulting in oscillations around the equilibrium point, in a way that will be discussed more fully in Chapter 5. These oscillations can in fact be seen in the upper figure of Fig. 3.5. This example shows that simple plots of Δn against n can bring out interesting features of the data. More elaborate methods, using $\Delta \log n$, are discussed in Chapter 5. Before coming to these, the discussion of the theory of population dynamics must be extended to show what can be done to build and test mathematical models of population systems.

4 Mathematical models: continuous cultures and fishery dynamics

The theory of regulated populations developed in the last Chapter is a general one and can be applied to any population. The extension and application of the theory to one particular population requires a much more precise formulation of the functions for births, deaths, immigrants and emigrants. In this Chapter, two well developed population models will be displayed. The first, that of continuous cultures, is the fullest mathematical model for any population system, covering all the main factors affecting the population. Not surprisingly, it is a rather simple, and entirely artificial system, occuring only in laboratories and factories. There are quite close analogies between continuous cultures and some natural systems, and the way this model has been derived shows how others could be for use in the analysis of natural populations.

The second model is one developed for predicting changes of quantity and variations in the weight of fish caught by trawling at sea. It shows the practical and economic side of population dynamics. The correct management of any population requires a knowledge of what happens when its members are harvested, whether fully or partially, whether selectively or not. A fish population is complex in a way not considered in the previous Chapters. The fish vary not only in age, but also in weight, and the prediction of the total weight that will be caught is an important aim of the fishery biologist. The complexity of marine communities is such that the model only attempts to study some important variables in a fishery, while the continuous culture equations are an attempt to study all the important variables. On the other hand, the fisheries work shows more clearly the cycle of observation; model building; testing by new observations and experiments; and model refining. It is the most extensively tested model in population dynamics.

Continuous cultures

Continuous cultures are populations to which material is added and from which material, including part of the population, is taken away continuously. They can be of any scale, from very small ones in laboratory glassware of 100 cm^3 or so upwards, though pilot plants, to large factory systems used for producing marketable products. Their importance in population analysis is that the equations describing the changes that are to be expected in the culture have been developed to a more satisfactory state than for any other population system.

These equations are built up from terms that have a clear biological and physical meaning, and which can be tested.

Normally, a continuous culture settles down quickly to a steady state, in which the population density and all the constituents of the culture remain just about constant. This is very useful, both industrially and in biochemical experiments, and the equations that are developed in this Chapter describe this steady state. It is interesting to remember, when reading the dispute about whether natural

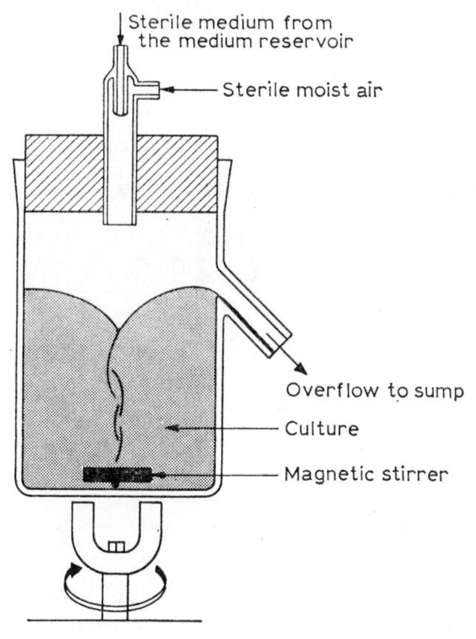

Fig. 4.1 A simple continuous culture apparatus (after Tempest 1970a).

populations can be described as being in a steady state, that continuous cultures are normally quite clearly so.

In principle, almost any organism can be cultured continuously, or at least with such small variations in flow rate that it can be considered continuous. Both in 1952 and 1963, F. E. Smith recommended such cultures as being best for experimental laboratory populations, but, except for some unpublished work, only populations of microorganisms have been studied this way. With these and particularly with autotrophic bacteria and with yeasts, the essential conditions for successful continuous cultures are quite simple. The organisms are grown in liquid media, solutions of salts and simple organic compounds, in a closed container. The volume of the culture is kept constant by an overflow system. The culture is kept well stirred and at a constant temperature and sterile medium is fed continuously to the culture from a reservoir. A simple system is shown in Fig. 4.1. For large-scale work and for fast growth rates, there are various technical difficulties discussed by Malek and Fencl (1966) such as ensuring sufficient aeration and a constant pH, but satisfactory cultures of a few hundred cubic centimetres growing fairly slowly are easily set up in the laboratory.

The basic equations described in the system are derived easily from the fundamental equation of population dynamics

$$d \ln n / dt = b - d + i - e$$

If the conditions in the culture are satisfactory, there need be no death rate, and as the incoming medium is sterile, there is no immigration. The equilibrium is reached when the rate of increase of the microbes is equal to the rate of loss over the overflow. This system will be

$$d \ln n / dt = +f_1(n) - f_2(n),$$

where $f_1(n)$ is the function describing population growth, by binary fission in most microorganisms, and $f_2(n)$ the rate of loss of the population over the overflow. That is,

$$f_1(n) = b, \; f_2(n) = e$$

For a well stirred culture the rate of loss is related to the flow rate. If the volume is V and the flow rate is f, then the dilution can be defined as

$D = f / V$ and the form needed for $f_2(n)$, the loss, is

$$d \ln n / dt = -D \qquad (4.1)$$

This loss must be balanced, at equilibrium, by the growth of the culture. A continuous culture can be kept in equilibrium either by adjusting the flow rate to the growth rate, or arranging the conditions so that the growth rate adjusts itself to the flow rate. The first type is called a turbidostat. A measure of the population density is fed back to the pump controlling the flow rate, so that the density is kept constant. There are technical difficulties in this, and the system is not particularly interesting to the population analyst, eacept that it can be made to record the growth rate of an organism under a variety of conditions. It corresponds to Fig. 3.2 E, except that a variable emigration rate replaces a variable death rate.

The other system, in which the state of $b = e$ is achieved by compelling b to vary, is clearly a system with density dependent population control. It corresponds to Fig. 3.2 F. It is usually called a chemostat, and the theory of the system was developed by Monod (1950) and Novick and Szilard (1950), and the clearest exposition of their theory is that of Herbert *et al.* (1956), or Tempest (1970b).

Theory of the chemostat

The chemostat depends on feeding the culture with a medium in which one known substance only is limiting the population growth. It might be glucose for example if this were the sole source of carbon for the microorganism. The equations in fact involve two variables, the concentration or population density of the microorganism, and the concentration of the limiting substrate. As this latter is naturally measured as a weight (either in the whole culture or per unit volume), it is convenient to talk in terms of the weight of the microorganisms per unit volume too, though it is usually fairly easy to convert the weight to numbers if one wants to.

For reference, it is probably most convenient to use the symbols of Herbert *et al.*, as these are now widely used. The variables needed are:

x the weight of microorganisms in the culture
s The weight of the limiting substrate in the culture
(the two primary variables measured as concentrations, i.e. weight per unit volume.)

μ the specific growth rate of x, equivalent to b, or $f_1(x)$

μ_m the maximum growth rate of the organisms under the conditions set
D the dilution rate, already defined
s_R the concentration of the limiting substrate in the incoming medium from the medium reservoir
K_s a constant relating μ to s
Y a constant relating the yield of x from s.

There is experimental evidence that the growth rate of microorganisms when grown with a single limiting substrate of this sort can be described by the equation

$$d \ln x / dt = \mu = \mu_m \left\{ \frac{s}{K_s + s} \right\} \tag{4.2}$$

which gives a curve like that of Fig. 4.2.

This is the form of the Michaelis equation in enzyme kinetics. It is clearly the sort of equation needed, in that when $s=0$, $\mu=0$, with no substrate there is no growth, and there is a maximum growth rate under these conditions when the substrate is no longer limiting. The value of K_s is defined as the concentration of s for which $\mu=\frac{1}{2}\mu_m$, and the concentration of the limiting substrate in a continuous culture is frequently not greatly different from K_s.

Putting together the expressions for growth and loss, equations *4.1* and *4.2*, we get

$$\frac{d \ln x}{dt} = \mu_m \left\{ \frac{s}{K_s + s} \right\} - D \tag{4.3}$$

At equilibrium this must equal zero, and by manipulation of the right hand side it is clear that the concentration of the limiting substrate when the population of the organisms is steady, $\hat{s}$ say, is

$$\hat{s}=DK_s/(\mu_m-D) \tag{4.4}$$

It is perhaps a little surprising that having set the conditions for equilibrium of the microorganisms, an expression is found for the concentration of the limiting substrate. To find the concentration of the microorganisms, conversely, we consider the rate of change of s, which depends on three processes, the rate of

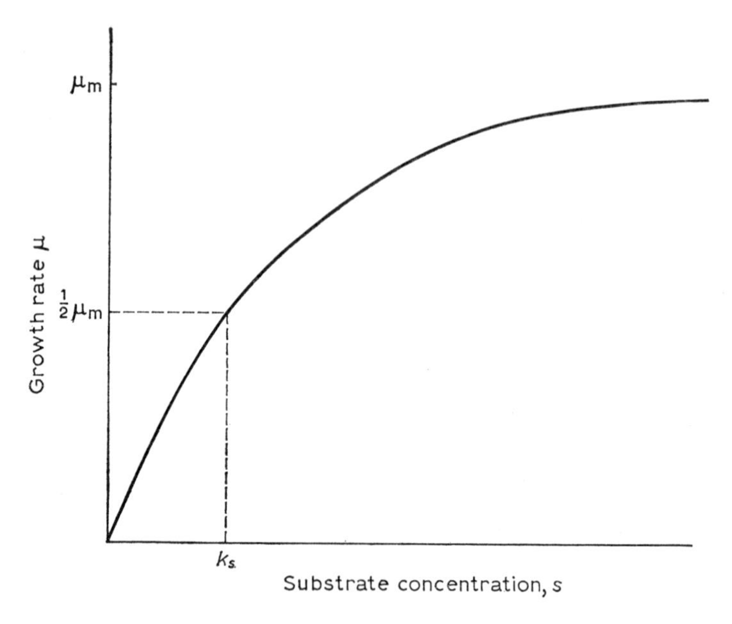

Fig. 4.2 The relation of growth rate to substrate concentration in equation (*4.2*). K_s is the substrate concentration when the growth rate is half of μ_m.

input from the reservoir, the rate of output over the overflow, and the rate of consumption of the substrate by the microorganisms. The first two of these processes are, quantitatively Ds_R and Ds. The third relates to the incorporation of molecules into the biomass of the microbes. For many media over a wide range of growth rates it has been found that:

weight of microbe formed = weight of substrate used × a constant.

The constant here is the yield constant, designated Y in the list above

$$Y=\frac{\text{weight of bacterium}}{\text{weight of substrate}}=\frac{\text{growth rate}}{\text{consumption rate}}$$

and so from this

$$\text{consumption rate} = \mu x / Y,$$

as we are considering the growth rate of the culture as a whole, and not specific growth rate of individual microorganisms. So, putting these three terms together,

$$ds/dt = Ds_R - Ds - \mu x / Y \tag{4.5}$$

and again, at equilibrium both sides equal zero. This gives

$$\mu x / Y = D(s_R - s)$$

and substituting for s from (*4.4*), and as $\mu = D$ at equilibrium,

$$x = Y \left\{ s_R - \frac{DK_s}{\mu_m - D} \right\} \tag{4.6}$$

Of these Y, μ_m and K_s are parameters of the biological system, while D and s_R are set by the scientist. The system described in Herbert *et al.* is *Aerobacter cloacae* growing in a fairly simple mixed salt solution with glycerol as the source of carbon. *A. cloacae* is a gram-negative rod bacterium, also known as *Enterobacter cloacae*. They made the carbon source the limiting substrate and found that:

$$\begin{aligned} Y &= 0.53 \\ s_R &= 2500 \text{ /lmg} \\ K_s &= 12.3 \text{ mg/l} \\ \mu_m &= 0./85\text{hr} \end{aligned}$$

Note that Y is less than 1 even though the bacterium consists of many compounds derived from glycerol. This is because glycerol is used as an energy source, and broken down to carbon dioxide as well. Note also, s_R is far larger than K_s. At the dilution rate of 0.5 per hour, that is, when half the volume is replaced each hour, the weight of bacteria in milligrams per litre is equal to

$$\begin{aligned} &0.53\,(2500 - 0.5 \times 12.3)/(0.85 - 0.5) \\ &= 1315 \\ &= 1.3 \text{ gm/litre or about } 1.3 \times 10^8 \text{ bacteria per cm}^3. \end{aligned}$$

Implications of the equations

These equations form the only reasonably satisfactory model for a population system with two components, in this case the bacterium and its limiting substrate. For one thing they relate to a system that works, unlike, say the classical predator/prey equations discussed in Chapter 14. Continuous cultures are commonly used on an industrial scale, such as in brewing beer. The model gives moderately good predictions of the behaviour of some, though certainly not of all, continuous cultures; but its attraction, and the reasons for its importance in

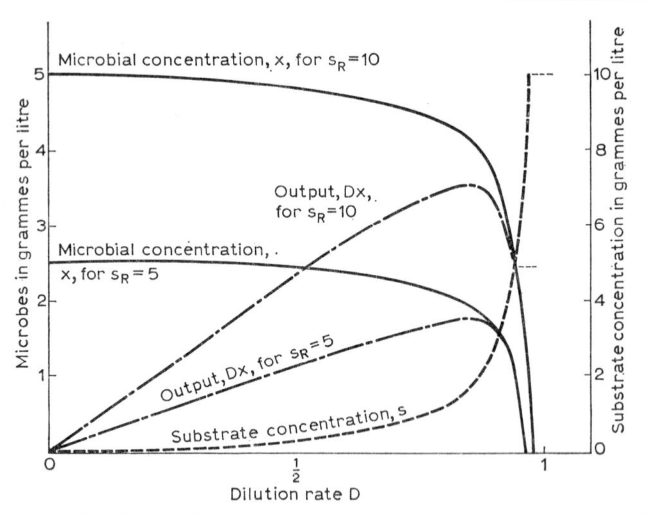

Fig. 4.3 Expected variations in the microbial population size, the output of microbes and the concentration of the limiting substrate in the culture with dilution rate, and at two concentrations of the limiting substrate in the reservoir. $\mu_m = 1.0\ \text{hr}^{-1}$, $Y = 0.5$, $K_s = 0.2$ g/l. s_R of 10 g/l and 5 g/l. Continuous line : microbial concentration, x. Dashed line : microbial output, Dx. Dotted line : substrate concentration, s, which is the same with both values of s_R. (after Herbert, Elsworth and Telling 1956).

showing how models may be used in population analysis, is that it builds up from sets of equations which are in themselves both reasonable and testable.

Equation 4,1 depends on homogeneous mixing in the culture, and so its validity is dependent on the nature of the culture vessel. For any particular culture vessel it can be tested. The other two main assumptions are biological: the relation of growth rate to substrate concentration in *4.2* and the expression for consumption rate in *4.5*. These are harder to test by themselves, because in any batch test the substrate concentration will vary. In a continuous culture both or some alternative, must apply and they can only be tested together. This can be done by altering the two variables which are under the experimenter's direct control, namely the dilution rate, D, and the concentration of the limiting substrate in the reservoir s_R. The effects that the equations predict for such changes are shown in Fig. 4.3. Shown there is the change in bacterial concentration, substrate concentration and bacterial output, for a system similar to that of Herbert *et al.* described above. It also shows the bacterial culture expected with s_R of 10 and 5 g/1. It will be seen that there is a maximum dilution rate, dependent to a slight extent on s_R. At faster dilution rates, the organisms should be washed out. It can be shown that the maximum dilution rate is

$$D_{max} = \mu_m s_R / (s_R + K_s) \qquad (4.7)$$

which is clearly slightly less than μ_m the maximum growth rate on the medium.

An interesting observation

The extent to which real continuous cultures depart from the theory described here is not very great, and is in the direction that might be expected. Herbert *et al.* (1956) showed that a 20 litre reaction vessel, with various side pockets for sampling, holding recording instruments and so on, was able to support populations at flow rates greater than that given by equation *4.7*. They ascribed this, reasonably, to a lower effective dilution rate in the side pockets. Postgate (1965) discusses the variation in the yield constant with different types of limiting substrates. Carbon sources generally behave in accordance with the theory. The largest discrepancy occurs with magnesium limitation: this sometimes produces appreciable numbers of dead cells in the culture, a further complication. Tempest (1970a) discusses the effects of Magnesium and Potassium limitation in some detail.

Biologically, a more interesting observation was made by Powell (1958) of the presence of a contaminant. He showed that if a new organism that depends on the same substrate is introduced into a continuous culture, whichever of the new and old organisms has the faster growth rate will displace the other. This is a simple case of competitive displacement, in a system with only one density-dependent controlling factor (Williamson 1957). As the conditions in the culture are of course normally arranged to suit the organism that the experimenter wishes to grow, continuous cultures often give less trouble with contaminants than batch cultures.

Powell's contaminant was found in a large reaction vessel with *Aerobacter cloacae* the bacterium used by Herbert *et al.* (1956). The contaminant was *Pseudomonas aeruginosa*, the bacterium used by Gause for growing Paramecium. Pseudomonas would scarcely grow on the medium provided for Aerobacter, and it appears that it was maintaining itself on some unidentified substance produced by Aerobacter, a sort of primitive food chain. A comparable situation may occur in plankton (Lucas 1961) with the metabolites of one organism affecting the growth of others, but the population densities are far less. The extent to which food chains consist, in this way, of organisms using the surplus production of other populations, and the extent, conversely, to which a population has a detrimental effect on those it feeds on, will be considered in Chapters 13 and 14.

Continuous cultures have been developed mostly by biochemists and industrial microbiologists, but they are good population systems. One such will be mentioned in Chapter 14. The importance of the system is that it shows that a fairly simple, biologically reasonable, testable and usable population model can be made.

The dynamics of exploited fish populations

Another model developed because of its economic importance is that of

Beverton and Holt (1957) for the dynamics of exploited fish populations. The fish they were particularly concerned with were those caught by trawling in the North Sea. Trawling involves dragging a large net by a powerful fishing boat along the botton of the sea. Fish living on or near the sea floor are caught in the net as it moves forward. Small fish can escape through the meshes, but large ones cannot and are, at the end of a trawling run, hauled inboard and eventually landed and sold. The price given for the fish of one species depends primarily on their weight, so the weight of the fish caught is the main variable that the fisheries biologist studies.

By the simple development of the fundamental equation of population dynamics, the change in the total weight of a fish stock is

$$\Delta S = G + R - F - M,$$

where G is the growth in weight, R the weight of new fish recruited to the population, F the mortality from fishing and M the 'natural mortality', which includes mortality from all other causes and the net balance of emigration and immigration of old fish. Beverton and Holt's work consists of the progressive development of suitable mathematical forms for each of these, and the development of the necessary techniques for estimating the parameters.

The first part of the work develops a simple model for the four factors. Growth is described by the von Bertalanffy growth curve. A simple, even naive, derivation of this is to say that the gain in weight is proportional to the surface of the organism, as all nutrients have to pass through a surface, while the loss of weight from respiration is proportional to the mass. This gives

$$d\,w\,/dt = al^2 - bl^3$$

where w is the weight, l the length (or other linear dimension) and a and b are constants. As w is proportional to l^3, this reduces in differential form to

$$d\,l\;dt = g - hl, \text{ where } g \text{ and } h \text{ are further constants.}$$

This can be integrated to show that the curve of weight against time will be a sigmoid one with an asymptote W_∞ for very old fish, and an inflection at about $0.3W_\infty$. The equation is

$$w_t = W_\infty(l - e^{-Kt})^3, \qquad (4.8)$$

where K is a constant.

Recruitment in numbers is assumed to be constant in the simple theory. This in effect regulates the system, as the number of recruits is always the same irrespective of the size of the adult population. This assumption is reasonably accurate for commercial fish populations. The fish when first recruited suffer only from natural mortality, because they escape through the trawl nets, but from a certain critical age, and so weight, suffer both fishing and natural mortality. Both forms of mortality are assumed to act randomly over the whole population, so the changes from these, in terms of number of fish are

$$d \ln N / dt = -M$$

for the small fish and

$$d \ln N / dt = -(F+M)$$

for the larger ones, F and M being the parameters of fishing and natural mortality.

It can be shown that at equilibrium, the total catch from a fishery of all ages in one period is the same as the total catch of one cohort of the population over it life span. For such a cohort let the numbers recruiting to the stock be R, then the number left at the age at which they are first caught, t_1 is

$$N_{t1} = R \exp\{-Mt_1\} = R' \text{ say}$$

and the number left at older ages is

$$N_t = R'.\exp.\{-(F+M)(t-t_1)\} \qquad (4.9)$$

The total weight of the population at time t is $N_t.w_t$, obtained by multiplying the right hand sides of *4.8* and *4.9*. The rate that weight is caught is dY/dt, the rate of change of yield, and this is $F.N_t.w_t$. The total yield from the cohort is found by integrating dY/dt. The integral is a rather formidable looking expression, largely because of the interaction of the cubic term of *4.8* with the other terms, and need not be written in full here.

Elaborations of the theory

The remainder of Beverton and Holt's book consists of working out the consequences of this expression for the yield of the fish, and its elaborations. In their second section they consider mathematical effects of using more complicated forms for recruitment, for natural mortality, for fishing mortality, and for growth. They are particularly concerned with variation of the parameters with the density of the population. They also consider variations in the population in space, and some aspects of mixed populations, though they are only considering the yield in weight from mixed populations. Their theory here seems a little limited compared with the theories that will be expounded in Chapters 13 and 14.

The third section of Beverton and Holt develops the next important aspect of model building, the estimation of the parameters. They consider in detail the estimation of these for each of the four factors, and this section is a most thorough exposition of the problems and excitements of developing and testing a mathematical model from real data. In their fourth section they come to the application of their model, of which the central feature is the eumetric fishing curve shown in Fig. 4.4. This shows the variation in yield, for variations in fishing intensity, that is to say in F, and for variations in mesh size. The variation in mesh size affects the youngest age at which fish are caught, that is to say the time t_1 used in deriving

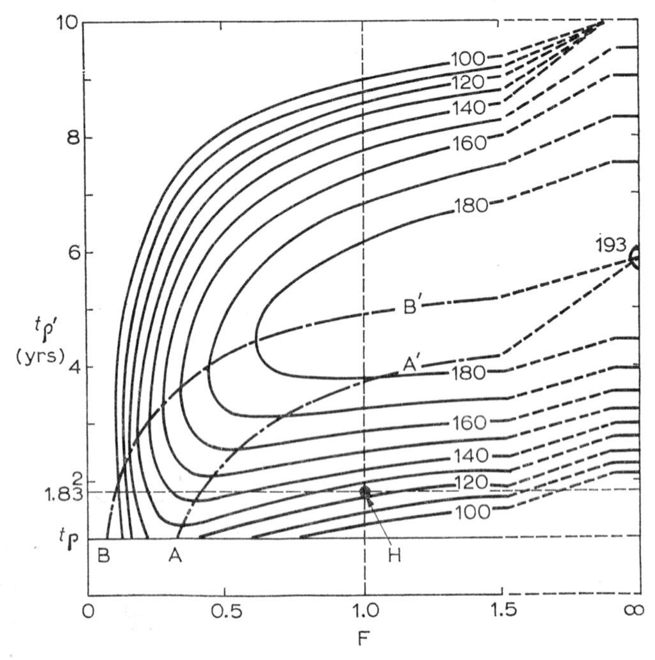

Fig. 4.4 Yield isopleth diagram for haddock. The contours show the yield in grams per fish recruiting to the bottom-living stock, at different values of F, the fishing intensity, and t_{ρ}', the age the fish first become large enough to be caught by the trawl. t'_{ρ} is therefore an index of the mesh size of the trawl. Curve AA′ gives the optimum fishing intensity for a given mesh size. Curve BB′, the eumetric fishing curve, the optimum mesh size for a given fishing intensity. The point H indicates the position of the North Sea haddock fishery in the 1930's.

4,9. The curves AA′ and BB′ show respectively the optimum fishing intensity for a given mesh size and the optimum mesh size for a given fishing intensity. For an international fishery like that in the North Sea, it is far easier to reach agreement on the minimum mesh size to be used, than to agree on the total number of boats fishing, which is the factor that affects the fishing intensity.

Continuous cultures and the dynamics of fisheries show two ways in which mathematical models can be developed, and show the sort of parameters which are derived and which need to be estimated. They show that it is possible to build models that are biologically reasonable, fit the observations, and are testable.

5 The analysis of changes in single species populations

The fundamental equation of population dynamics, which was discussed in Chapter 3, is

$$d\ ln\ n/dt = b - d + i - e$$

The corresponding fundamental task in the analysis of populations must be to find the relationship of the four terms on the right, and their component parts, both to variations in the population's density and to changes in the physical and biological circumstances of the population. In attempting to analyse a population this way, I consider first in what form the population data should be presented, and in what way the data should be compared with other variables.

There are four population variables which look promising. These are the population size (n); the change of population size ($n_{i+1} - n_i$ or Δn); the logarithm of the population size ($ln\ n$ or $\log n$); and the change in the logarithm of the population size ($\log n_{i+1} - \log n_i$ or $\Delta \log n$ or $\Delta\ ln\ n$). In Chapter 1, it was shown that the logarithm of the population size was often a more useful first plot than the population size itself. In Chapter 3 there was an example using Δn, though $\Delta \log n$ could equally well have been used, while for the continuous culture system in Chapter 4 the primary variable was $d \log x/dt$, the rate of change of the logarithm of the weight of the microbial population. So the theory, laboratory experiments, and observations on natural populations all suggest that the use of the change of the logarithmic populations size may be the most useful first statistic to consider.

There are other considerations that lead to the same conclusion. Consider a particular cohort of organisms at one stage in their life history, i.e. a set of individuals of the same species that were born at the same time, and then consider the change to the number still alive at some later time, this can be expressed in a number of ways. We can talk about the mortality from $n_1 - n_2$: that $x\%$ of n_1 has died. Mathematically this is

$$\frac{n_2 - n_1}{n_1} \times 100,$$

or logarithmically it is $\log (n_2 - n_1) - \log n_1 + \log 100$. Alternatively we can talk about the survival, that $y\%$ of n_1 survive. This mathematically is $n_2/n_1 \times 100$ or in logarithmic form $\log n_2 - \log n_1 + \log 100$. When building up the basic Leslie matrix in Chapter 2, it is the survival of each cohort in each time interval that forms the elements in the sub-diagonal. Lack (1954) has shown that for many bird populations there is a constant survival from year to year, once the birds become adult. The same seems to be true of many fish populations, once the individual members of the population become big enough to be caught, and thereafter up to an age at which the population is usually so thin that there are not enough to be worth considering. This is discussed by Beverton and Holt (1957), and also by Holt (1962). For a population where one can talk of new born, immature and adults, the Leslie matrix takes the form

$$\begin{bmatrix} 0 & 0 & f \\ p_n & 0 & 0 \\ 0 & p_j & p_a \end{bmatrix} \quad \begin{bmatrix} b \\ y \\ a \end{bmatrix}$$

where f is the average fertility of the adult, and $p_n\, p_j\, p_a$ are the survival probabilities of the new born, young and adults respectively. For any particular cohort of adults we have

$$a_2 = p_a a_1$$

$$a_3 = p_a a_2 - p_a^2 a_1 \text{ and hence}$$

$$a_{i+1} = p_a^i a_1$$

That is to say, the survivals have to be multiplied together to produce the final expected number in the cohort. From this, it follows that the logarithms of the survival can be added together to produce the logarithm of the final population. The equation becomes

$$\log n_{i+1} = i.\log p_a + \log na_1.$$

The life table in such a case is of negative exponential form when the population is plotted arithmetically, and is a straight decreasing line when the population is plotted logarithmically. Exactly the same mathematical form was used for the rate of wash-out of microbes in the continuous culture equations in Chapter 4. This negative exponential form can be expected whenever the chance of an individual surviving depends only on the fact that it is a member of a particular population, and not in any way on the previous history of that individual.

So far, I've only discussed survival in successive time intervals. If there are two independent causes of mortality, acting simultaneously, in the same time interval, then the mathematics are much the same. If we have a population n_1, and if the expected number at time two because of the first mortality factor is $s.n_1$ and from the second mortality factor is $t.n_1$, where both s and t are probabilities of survival and so lie in the range from 0–1, then the expected population size when both mortalities are acting independently is $n_2 = s.t.n_1$. Again this becomes additive if put in logarithmic form:

$$\log n_2 = \log s + \log t + \log n_1.$$

So for mortalities in successive time intervals, or for mortalities acting independently in the same time interval, the system becomes additive provided the survivals are expressed as log survivals.

In any statistical analysis, it is convenient to be able to use the analysis of variance. This, as is well known, is a powerful and flexible technique, and can be used whenever effects are independent and additive, provided that they are not too non-normal in their distribution. The logarithm of the survival of the cohort, that is $\Delta ln\ n$ or $\Delta \log n$, is additive in this way, and so can usually be studied by the analysis of variance. Although this seems not to have been done, it is another strong reason for dealing with $\Delta \log n$ as the primary variable when trying to analyse the population. There are two fields in which $\Delta \log n$, or $d \log n/dt$, have been much used. These are fisheries and entomology.

Mortality in fisheries

In fisheries the numbers of the cohort can be studied by estimating the age of fish in a catch from their scales or sometimes from their otoliths. As has been discussed more fully in Chapter 4, one of the important things a fisheries scientist must know is the effect of catching the fish on the population. That is he needs to estimate the mortality from fishing. To do this he has to separate mortality into two components, fishing mortality and 'natural mortality'. Fishing and natural mortality are usually independent, even though they act during the same time between two different censuses of the population. The conventional notation in fisheries for this is

$$N_t = N_o e^{-zt} \text{ or } ln\ N_t = lnN_o - Zt$$

or for unit time,

$$\Delta lnN = F + M = Z = Z$$

where Z is the total mortality and can be split into two components, F and M. This gives $ln\ N_t = ln\ N_o - (F+M)t$, or for unit time, $\Delta\ ln\ N = F+M$. The type of basic data that are available in this sort of study are shown in the plot of the herring population in the north-west North Sea in Fig. 1.20 (page 17). This shows the history of the successive cohorts in the population, and the slope of the line shows the mortality from year to year of each cohort. One of the striking things about this diagram is that in the right-hand half of the graph there have been appreciable changes in the mortality rate in the population and these changes appear to have affected all age classes simultaneously. An important question in studying a fishery such as this is, how much of the variation in mortality is due to variations in natural mortality and how much in fishing mortality? For herring, this type of analysis is set out in Anon. (1970), a study of the stocks in the Norwegian Sea. There has been a drastic decline in the stock of adult fish, caused partly by a high exploitation rate, and partly by a lack of recruitment. There are also fisheries for young fish, and these are partly responsible for the lack of recruits to the adult stock. However, there are wide and so far unexplained variations in the strength of the year classes, and remarkable changes in the geographical location of the stocks from year to year. For the adult stock, natural mortality, M, has fluctuated around a figure of 0.16, but the fishing mortality, F, has increased from 0.05 in the 1950's to 0.27 in the late 1960's. It is scarcely surprising that it is recommended that the rate of fishing on the adult stocks should not increase, and that some of the fishing on the young stock should decrease.

k factor analysis

The other field in which the logarithm of survival has been much used is in entomology, particularly in the study of lepidoptera and other insects that attack the leaves of trees, and their hymenopterous and dipterous parasites. This sort of insect commonly has one generation a year, though some species have more. Consequently the survival that is studied is from stage to stage and not from year to year. It is possible to estimate the numbers in the different instars, which would normally be active at different times of the year, with reasonable accuracy. Such a study is shown in Fig. 1.17 (page 14), Klomp's study of the numbers of the stages of the pine looper moth (*Bupalus piniarius*) in the conifer forests at Veluwe in the Netherlands. Figure 1.17 shows the logarithm of the population size, and so the logarithm of survival is given by the difference between successive points vertically on this plot. These are shown plotted as variables in their own right in Fig. 5.1. The successive log survivals are labelled by a series of variables of type k_i, k_1 being the survival from eggs to the first instar. This notation and method of analysing populations is due to Varley and Gradwell (1960). As originally defined by Varley and Gradwell, the successive survivals are labelled $k_1 \ldots k_n$ with $\sum_1^n k_i = K$. The graph of k_i shows at a glance which stage of the life history has been most variable, and one can pick out if there has been different survivals which are correlated with one another. These impressions can be tested statistically, by calculating the variance of any individual k and the correlations between the separate k's. It is also possible to test for density dependence at any stage by testing the relationship of a k_j with the n_j, that is to say the numbers at the beginning of the survival period involved in k_j. At first sight this might appear statistically invalid as

$$k_j = \log n_j - \log n_{j+1},$$

and so to compare k_j with $\log n_j$ is comparing a figure with itself to some extent. However the comparison is statistically equivalent to comparing $\log n_j$ to $\log n_{j+1}$ which is valid, though it is usually a less convenient form and one in which it is less easy to see density relationships. The same criticism can be levelled against the use of Δq for the change in gene frequency, which is dealt with in the next

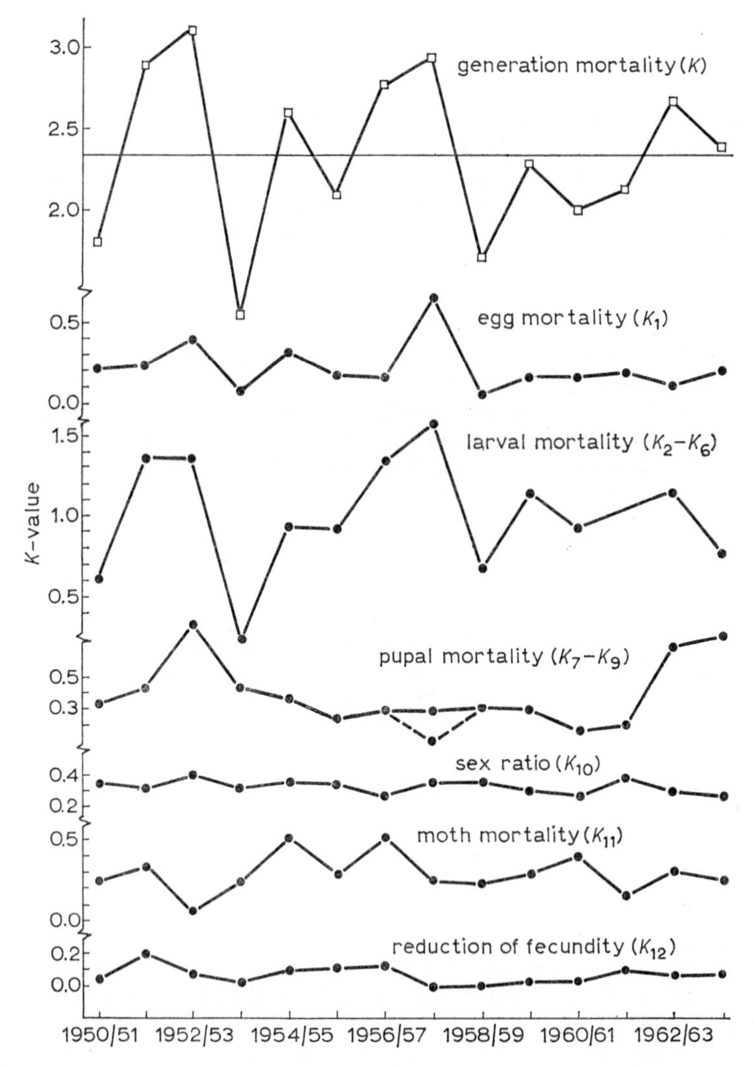

Fig. 5.1 *k* factors (the negatives of the $\log_{10}$'s of the survivals) of *Bupalus piniarius* at de Hoge Veluwe from 1950 to 1964. Note the various scales to the left. These *k* factors are derived from the data in Fig. 1.17 (from Klomp 1966).

Chapter. There the comparison of q_i with q_{i+1} is valid, but the comparison of Δq with q is more convenient.

It has been claimed by Eberhardt (1970) and Watt (1968) that the comparison of Δn with n and related techniques are misleading because, with these tests, random numbers appear to be controlled populations. Random numbers are only random for certain purposes, and these do not include immitating random walks. A set of random numbers from a table considered as a population has a strictly limited, rectangular distribution, and represents a strictly, if unusually, controlled population. If the range is say, 00 to 99, then this can represent a population, in a test correlating Δn with n, varying from 1200 to 1299 with a mean of 1249.5. This is a very restricted range, and it would be disturbing if such a set of numbers did not appear to be controlled with appropriate tests. This point was made by several people at the Advanced Study Institute in September 1970, particularly clearly by Reddingius (See Den Boer and Gradwell 1971).

Insect populations of this sort have an interesting and important feature which is not found so commonly, or at any rate so strikingly, with other organisms. The insect parasitoids (hymenoptera and diptera) that attack the insect herbivores, mostly lepidoptera, have a generation time that is the same length as that of their hosts. So if a hymenopterous parasitoid adult female attacks the second instar larva of a particular species of lepidopteran, the development of that hymenopteran is timed so that the next generation of females is flying at the time the next second instar of the lepidopteran are available. This automatically introduces a lag of one generation into the population effects. The first important full study of this was by Nicholson (1932), and the population pattern to be expected, if the attacks of the parasitoids are density dependent, was named delayed density dependent by Varley (1947). Delayed density dependence, like other population effects, can be easily studied by *k* factor analysis. Some illustrations of these techniques are given in Fig. 5.2, from Southwood (1966).

Mortality analysis used by fisheries biologists, and the *k* factor analysis of Varley and Gradwell are in fact mathematically equivalent. Both depend on being able to estimate the numbers of one cohort at successive stages in its life history. There is a trivial difference between them. Fisheries biologists work with natural logarithms, Varley works with logarithms to the base 10, but one can readily be converted into the other by a constant. The major contribution of *k* factor analysis to the analysis of population size is to show which forms of mortality are large and variable and which small. The analysis of density dependent effects whether direct or delayed have been more disappointing (Klomp 1966, Broadhead and Wapshere 1966, Varley and Gradwell 1970) but Krebs (1970) shows the importance of clutch size and hatching success in regulating the Marley population of the great tit. It would be natural to relate the variability of any particular *k* to changes in the physical environment. Varley and Gradwell (1962) are inhibited from doing this for fear of finding false and misleading correlations. This is certainly an ever-present danger when using correlations to study any non-experimental situation, but progress is nevertheless possible as will be discussed more fully in Chapter 16 in relation to the principal component analysis of marine plankton. The point made there is that it is possible to generate hypotheses, and so develop tests for hypo-

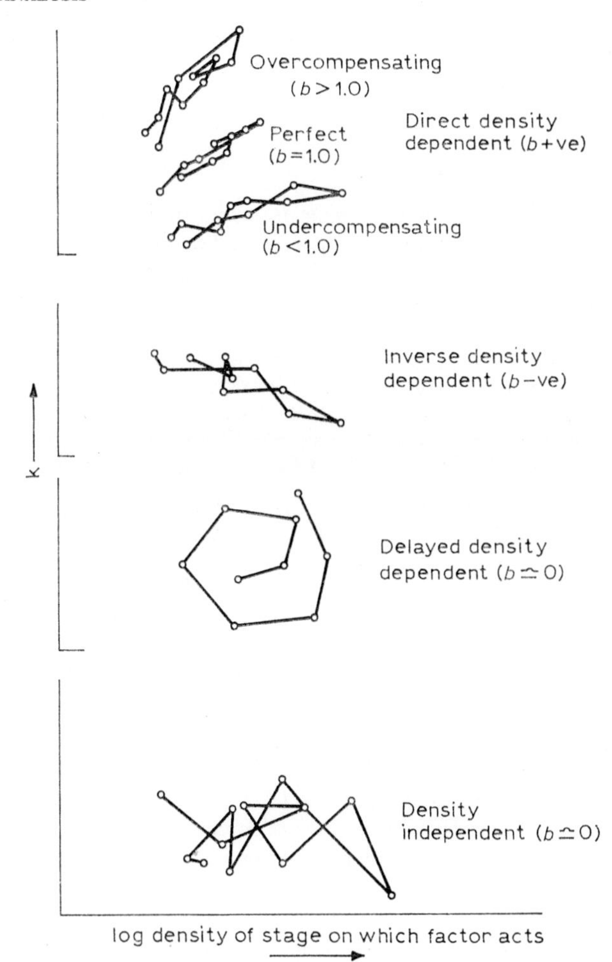

Fig. 5.2 Time sequence plots showing the relation of *k* factors to population density with different density relationships. The *b*'s are the slopes of the relation of *k* on *ln n*. (from Southwood 1966).

theses, from correlation analyses. It is not possible to give rigorous demonstrations of relationships by such analyses.

So far we have only discussed mortality. However, fertility, that is to say the top line of the Leslie matrix, can also be studied conveniently by the logarithmic transformation. If each female in a population has an expectation of laying x eggs, then the expected number of eggs is x times the number of females. Put in logarithmic form this is $\log x + \log$ (number of females) or in other words the fertility of the female can be considered as yet another *k* in the Varley and Gradwell notation. It differs from the other *k*'s in having an opposite sign, because it relates to an increase in the population rather than a decrease.

Overlapping generations

The analyses so far refer to organisms in which it is possible to follow the history of single cohorts throughout their lives. This is not always possible, and we may have a situation as in the matrix on page 52, where one can distinguish the stages up to adult, but not different ages of adults. Such a case is shown in

Table 5.1 Cockles in areas A, B and C, Llanrhidion Sands, per m^2, from samples. Left hand columns first year cockles, right hand columns older cockles. Data from Hancock and Urquhart (1965)

	1958/9		1959/60		1960/61		1961/62	
November	1973	133	2386	612	2926	615	3176	821
February	814	69	1306	89	1134	54	1829	95
Apr/May	581	93	1150	93	1512	57	1181	64
July/Aug	557	63	793	88	986	47	1159	44
November	577	35	556	59	784	37	688	35

Table 5.1, which is the data from Hancock and Urquhart (1965) on the numbers of cockles *Cardium edule* on the Llanrhidian sands in South Wales. A *k* factor analysis of this data, which I leave to the reader, shows that the major mortality is between November and April at the beginning of the second year and that the mortality is density dependent. Hancock (1971) showed that this mortality was largely due to oystercatchers (*Haematopus ostralegus*). A *k* factor analysis can reasonably be tried whenever there are data which are collected at shorter intervals than the generation. An example of this is shown in Fig. 5.3, which is the change of log population size for the Canadian lynx data on Fig. 1.7 (page 6). This figure brings out two points which are certainly not so clear in the plot of the population size. The first is that the population is scarcely ever stationary, though in the original plot it appeared to be so at least at the bottom of the troughs. The second is that the cycles are quite distinctly asymmetrical. The average rate of decrease is quite appreciably faster than the average rate of increase. This was noted by Moran (1953) when studying the residue from figures fitted by a serial regression, but in that form it appeared to be a small and by no means clear effect. Plotted this way it is a large effect, and suggests a biological explanation. The rate of increase averages about times two. Small cats typically have a litter of four kittens of which two would be male and two would be female. Consequently one litter a year would produce an increase of about times two. So the increasing part of the lynx cycle either implies that all females are breeding at about a rate of one litter a year, or alternatively that if successful breeders can breed faster

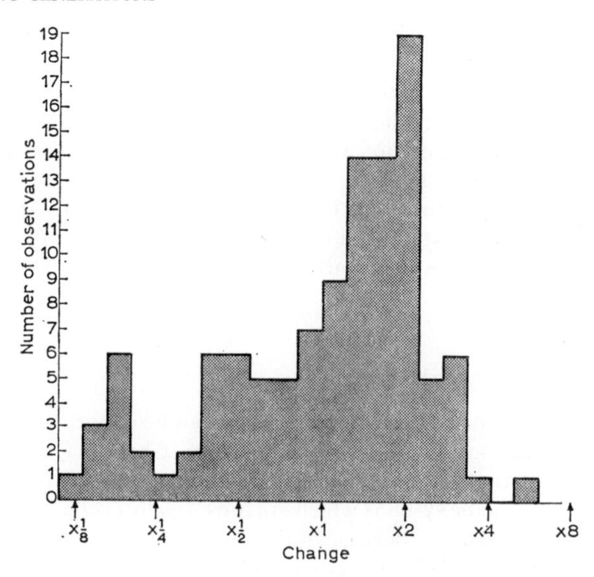

Fig. 5.3 The distribution of the change of the logarithm of the number of Lynx pelts from the Mackenzie district of Canada. The change is shown as a multiplicative factor for convenience.

than this, this is compensated by other females who are not breeding. The former hypothesis looks somewhat more likely. On the other hand the decrease must be related to the mortality rate under starvation. Elton and Nicholson (1942) quoted evidence that the lynx is almost entirely dependent on the snowshoe hare (*Lepus americanus*) for food and during the decrease period most of the lynx must be finding it extremely difficult to get any food at all. An interesting question that follows from that, is why, with such strong selection repeated so often, has there not been an evolution to a more diversified diet?

Varley and Gradwell's analysis was put forward independently of Morris's (1963) key factor analysis, even though they have the same names. Morris's analysis relates to the comparison of counts one generation apart. This should relate to the Σk, including the fertilities, and should normally include within each figure analysed a density regulating component. So it can be expected that each population size will relate to the physical conditions in that generation, and that the change in population size from generation to generation will correspond only to the change in physical factors. From this way of looking at things, key factor analysis should not work, and indeed Southwood (1967) finds that it does not work if the population is not undergoing a considerable trend. Nevertheless Morris found it useful. The reason for this apparent anomaly appears to be the delayed density effect that occurs when insect parasitoids are important. A fuller discussion of parasite/host relationships is given in Chapters 13 and 14, but here one can say that a log survival analysis will frequently be of interest whenever the analysis relates to events shorter than a generation time or, alternatively, when it relates to changes between generations and where there are important delayed effects from one generation to another. It can not be expected to work when the interval is one generation and the change from generation to generation is a stochastic markov process of the first order, that it to say, not carrying over for more than one step. Morris' technique is also criticized by Maelzer (1970) and St. Amant (1970), but they do not discuss k factor analysis or Moran diagrams or the well known difference between a functional relationship and a regression.

Oscillations and Moran diagrams

One particular situation in which delayed effects can be important is where there are cycles in population number. As was shown in Chapter 1, such cycles undoubtedly occur, even though they are fairly rare. As will be shown in Chapter 14, it is possible that some systems which are intrinsically cyclical may not appear to be so on a simple analysis, because of the disturbing affect of random factors. Moran (1949, 1950, 1952, 1953, 1954) has made a detailed study of this. He points out that when any serial correlation of a series is significantly negative, then one has an oscillatory situation.* When there is a significant negative serial correlation, of any order, this means that a population which is above its mean can be expected to be below its mean with the delay indicated by the order of the serial correlation coefficient, and vice versa. With this statistic, Moran showed that there were significant oscillatory tendencies not only in the Canadian lynx, where they are exceedingly clear, but also in some game bird populations, though not in others.

Moran also suggested a simple plot for studying whether or not there are oscillations. This has been extended by Ricker (1954a and b, 1958) to form what he calls a reproductive curve. This is a plot of a number at one time (n_i) against the numbers at the time before (n_{i-1}), and so in a simple form is a scatter diagram indicating whether or not there is a significant first serial correlation coefficient. In data with small standard errors though the dots will not form a random scatter but will form a line. Figure 5.4 shows three different ways of studying this. One is the plot of the population against time, the second is a Moran plot, while the third is a plot of Δn against n. Moran plots can be made either by plotting n_i against n_{i+1} or by plotting Δn_i against n_i, or in either or these two ways but using logarithms. Mixed plots of arithmetic values and logarithmic ones are not helpful. The relation of these plots to oscillation is most easily seen in the delta plots. Here if the line has a slope between 0 and -1, from horizontal up to 45°, the population will approach its equilibrium directly. For slopes between

* Serial correlations of order s is the correlation between the numbers of a series and other members s steps before, and is a correlation of x_i with x_{i-s}

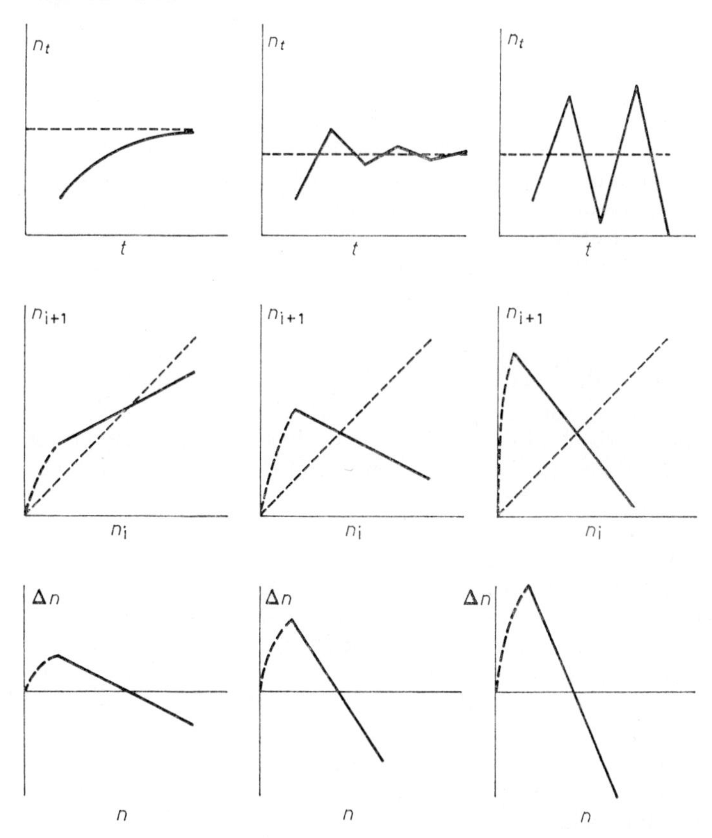

Fig. 5.4 Moran diagrams and related figures. The top line shows the change of population size with time with, left, a smooth approach to an equilibrium, centre, approach by a damped oscillation, and right, an increasing oscillation. The centre line shows the corresponding Moran diagrams, of plots of the population at one time against the population at the next time. The bottom line are the equivalent $\Delta n{:}n$ diagrams with, left, slope $<$-1, centre, slope $>$-1, $<$-2, and right, slope $>$-2. In logarithmic form (Δ *ln n*: *ln n*) the bottom line of diagrams are equivalent to the top diagram of Fig. 5.2.

—1 and —2, the approach will be through damped oscillations. At a slope of —2 there will be stable oscillations and for a slope greater than —2 there will be increasing oscillations. These are shown in Fig. 5.4. They can be compared with Fig. 3.5 for *Paramecium aurelia*, where it appears that the appropriate slope is somewhere near —2, and that there is something like a stable oscillation.

Figure 3.1 is an example of the use of a Moran plot, and it is clear that although the figures may fall around a line there is, as one would inevitably expect with real data some appreciable scatter. Another point that is not mentioned in the literature, is that Moran plots only work well if all the important delay is of exactly one generation. This of course is what would be expected in the insect parasite/host case, but not necessarily what would be expected in other cases. Plotting the lynx data in the Moran form for instance merely produces points that go round and round in closed curves, without generating any of the nice forms shown in Fig. 5.4. This is because the delay is of more than one step, and is in fact up to five steps. In theory this could be handled by using plots of an appropriate number of dimensions, but in practice, with graph paper being what it is, such plots are not convenient.

Putwain, Machin and Harper (1968) test the regression of n_{i+1} on n_i, and suggest that if this regression is significantly curvilinear, then there is evidence of density dependence. This technique is related to that of making a Moran plot, and has the advantage of using a standard statistical technique, but apart from this, it seems rather less flexible. What is needed is the functional relationship between the regression of n_{i+1} and n_i, and a regression line will only approximate to this if the variance of the points about the line is small, as the relationship will fall between the regression of n_{i+1} on n_i and the regression of n_i on n_{i+1} (see for instance, Sprent 1969).

The basic theory of population dynamics, the few laboratory experiments and the observation on natural populations discussed so far lead to a form of analysis based primarily on the change of logarithm of population size. This allows both the detection of density dependent effects, and the building of mathematical models for the action of population on itself. Further observations, and preferably experiments if these are possible, would lead to full models such as those of continuous culture. So far though, the effect of genetic variation within the population has not been considered, nor has the effect of populations of other species, and the analysis of the effect of physical factors on the population size has been treated cursorily. These will be considered in the remaining three Sections of the book.

II Genetic variability in populations

6 Simple population genetics of one locus

In most species each individual is genetically distinct. Its total array of alleles at all the loci of its genotype produce a set which is, as far as we know, different from that of any other individual of that species, and of course even more so of any other species. The most obvious manifestation of this is that it is quite easy to recognize individuals of a species. We know this for our own species, and those who work with other species find that individuals are soon recognisable. Some of this difference is environmental. It is possible to distinguish between identical twins, even though their genotypes are exactly the same. And there are species that produce clonally, either all the time, as apparently in some microorganisms and in apomictic plants, or for appreciable sections of their life history, such as the summer forms of Cladocera and Aphids. So for most organisms the implicit assumption of most of the discussion of Section I, that the differences between individuals are either specific differences or differences in age of the individuals, is not sufficient, we must also take into account the genetic differences between individuals. Individuals are genetically different, and conversely genes occur in genotypes, which occur in phenotypes, which are individuals. Neither population ecology nor population genetics is complete without the other. As Lewontin (1968a) says 'no population ecology can succeed that is not also population genetics'. Biologists must always remember that many phenomena, which at first sight seem to reflect fundamental laws of biology, may in fact be consequential effects of natural selection. This point is argued most cogently by Delbrück (1940). This is particularly important for population ecologists when they consider the interactions of two or more species, interactions such as competition or the interaction between the predator and its prey. These interactions will be considered in Section III.

Population ecology therefore must consider population genetics. The converse is also true. Genes occur in individuals, and therefore the fundamental equation of population dynamics

$$\Delta n = b + i - d - e$$

applies when n is the number of genes of a particular sort as well as when it refers to the number of individuals in the population. This point is usually obscure in discussions on population genetics, and there is a good reason for this. The mathematics of genetical systems are really quite difficult. In this Chapter, only the population genetics of two alleles at one locus will be discussed. Even elaborations considering several alleles at one locus, or alternatively, two alleles at each of two loci (Li 1967), produce very difficult mathematical problems that have not yet been completely solved. Further, all these mathematical systems deal with constant selective values. What is meant by constant selective value is described below, and its limitations are discussed in this Chapter and in the two next ones. So geneticists have usually worked with gene frequencies, ignoring the question of population size, and so of gene number. The consequences of this may often be unimportant and negligible, but in at least one problem, genetic load, it has led to an unnecessary dispute, as will be seen in Chapter 7.

As a first approximation, it is reasonable to assume that the factors regulating population size and the forces changing gene frequency are independent. In such a situation it is entirely legitimate to consider only the gene frequency. When there are two alleles in a population, say a_1, and a_2, if their frequencies are p and q, then as $p+q=1$, $p=1-q$. By dealing with the frequency, the problem is reduced to that of one variable only, and this is a sensible point to start.

Hardy-Weinberg Equilibrium

The basis of most discussions in population genetics, and the theorem that allows us to talk about the frequency of genes, even though genes occur in individuals and genotypes, is the Hardy-Weinberg Law, or as Manwell and Baker (1970) have it the CHW law, after Castle as well, who made some remarks that were relevant. In the simple case considered here, two alleles at one locus, the Hardy-Weinberg Law describes the frequencies that will be observed for the three genotypes a_1a_1, a_1a_2, and a_2a_2. The law states that after one round of random mating, the frequencies of the three genotypes will become p^2 to $2pq$ to q^2.

Unless some force is applied to the population system to change gene frequencies, the gene frequencies will stay at p and q, and the genotype frequencies at p^2: $2pq$: q^2 after the first round of random mating. This law applies to any number of alleles at any number of independent loci. When there is either sex linkage or linkage, the approach to the equilibrium genotype frequencies is slower, but still reached. Proofs of this law can be found in profusion in any standard text book of population genetics, such as Li (1955a). Here it will be shown to work using a matrix system, which is a development of the Leslie matrices of

Chapter 2, not just for a different type of demonstration, but more particularly to introduce the possibility of handling genetic systems by matrix methods.

The genetic random mating matrix for three genotypes composed of two alleles at one locus is as follows:

$$\begin{bmatrix} d+h & \frac{1}{2}(d+h) & 0 \\ r+h & \frac{1}{2} & d+h \\ 0 & \frac{1}{2}(r+h) & r+h \end{bmatrix} \begin{bmatrix} d \\ 2h \\ r \end{bmatrix} \qquad (6.1)$$

The column vector gives the frequencies of the three genotypes: d is the frequency of homozygote a_1a_1, $2h$ is a frequency of the heterozygote a_1a_2, and r is a frequency of the other homozygote a_2a_2. Because the Hardy-Weinberg equilibrium gives a frequency of heterozygotes of $2pq$, it is frequently more convenient mathematically to use $2h$, or a corresponding term, for the frequency of heterozygotes. The elements of the matrix come directly from the assumption of random mating, and can be said to define what is meant by random mating. The top left hand element of the matrix is $d+h$. This arises from two classes of matings. The probability that homozygote a_1a_1 will mate with another homozygote a_1a_1, is the frequency of that type in the population namely d. All the offspring of this mating will be of type a_1a_1, so the whole of this term d occurs in the top left hand corner. The rows of the matrix show the genotypes of the parental generation while the columns show the genotypes of the offspring generation. In the same way, homozygous a_1a_1 individuals will mate with heterozygote a_1a_2 individuals with a frequency $2h$. Half the offspring of such a mating, and so with frequency h, will be a_1a_1, while the other half, also with frequency h, will be a_1a_2. This explains the h in the top left hand element of the matrix and the h in the first column of the second row. All the other elements of the matrix can be derived in the same way. Multiplying the column vector by the matrix, we find that the resulting column vector is

$$[(d+h)^2 \quad 2(d+h)\times(r+h) \quad (r+h)^2]^T$$

The system can be written in terms of gene frequencies as

$$\begin{bmatrix} p & \frac{1}{2}p & 0 \\ q & \frac{1}{2} & p \\ 0 & \frac{1}{2}q & q \end{bmatrix} \qquad (6.2)$$

and the vector $\begin{bmatrix} d \\ 2h \\ r \end{bmatrix}$ goes to a vector $\begin{bmatrix} p(d+h) \\ dq+h+pr \\ q(r+h) \end{bmatrix} = \begin{bmatrix} p^2 \\ 2pq \\ q^2 \end{bmatrix}$

This random mating matrix can be adapted to any other mating system, or any type of selection or other genetic change (Williamson 1959, 1960).

The random mating matrix differs from the Leslie matrix in one important particular. The elements of the matrix are derived from the elements of the vector. It is still possible to derive the latent roots and vectors of the square matrix in *6.2*. The sum of the roots is evidently $p+q+\frac{1}{2}=1\frac{1}{2}$, from the sum of the elements of the principal diagonal. Because of the degeneracy implied by $p+q=1$, there will only be two non-zero roots. It is easy to show that the three roots are in fact 1, $\frac{1}{2}$ and 0. More surprisingly, the latent vector associated with the root of one half is $[+1, -3\ +2]^T$ which not only includes a negative term which can have no biological meaning, but which adds to zero. Further this vector carries the implication that $p+q=0$, contradicting the assumption that $p+q=1$, so altogether this root of $\frac{1}{2}$ has no biological meaning. The remaining root of 1 has the vector p^2, $2pq$, q^2, and the fact that this root is one shows that the population is in equilibrium.

Directional selection

The forces that can change the gene frequency in a population are very numerous but they can all be classified under three headings. These are migration, which is the movement of individuals in or out of the population; mutation, which is the direct chemical change of one of the alleles into another allele; and selection, which sums up all the remaining non-random changes in gene frequency. On top of these there will be random effects. Selection includes differential viability of different genotypes, and differential fertility. The simplest mathematical expression of selection is that of Haldane (1924) and his exposition of the change of frequency of a gene under simple selection is amongst the clearest available. Suppose there are two forms in the population, say A and B, with a frequency of A of 1 and B of p, where p is any number between zero and infinity and so B may be less frequent or more frequent than A. Then if the frequencies in the next generation are $1-s$ of A to p of B, s represents the force of selection against A. If selection is acting, then, in general, the frequencies of A and B will change, or in other words p in this expression will change. If s retains the same value irrespective of the value taken by p, then this is called a constant selective force. There are two things to note about this definition. The first that it is arbitrary although mathematically convenient. The second is that there has been no study of any natural population over a sufficient range of frequencies to say whether in fact any known selective force is in practice as constant as is required by this definition. There are a number of cases such as Kojima and Yarbrough's (1967) study of the esterase-6 frequency in an experimental population of *Drosophila melanogaster*, in which selection is quite definitely not constant. It would be useful to know how frequently the assumption of constant selective values is a reasonable approximation, and how frequently it is clearly wrong. This point is discussed further in Chapter 8.

Selection must act on individuals, that is on phenotypes (Milkman 1967, Mayr 1963). In dealing with selection on major genes of a diploid organism this is equivalent to saying that selection acts on genotypes (Rendel 1968). A typical selective system might be:

Genotype	a_1a_1	a_1a_2	a_2a_2
Selection	1	1	1-s

This form is for selection against the recessive, and the heterozygote and the dominant homozygote have the same selective value. In a system such as this, a_2 will eventually be eliminated from the population, and can only be re-introduced by the other two forces that affect gene frequency, namely migration and mutation. More generally, directional selection will lead to a balance of gene frequency between selection on the one hand and migration and mutation on the other. This tendency to eliminate a gene under directional selection will hold true even when the selective values are not constant, provided they do not change sign.

Selective elimination is a fairly slow process, even when the selection is strong. Haldane has tabulated some values, and some of these are given in Table 6.1.

Table 6.1 Number of generations needed for a change in the frequency of an allele under directional selection

(a) an allele of dominant phenotype				
Selection	1%	3%	10%	30%
Frequency change				
0.5 to 5%	237	79	24	8
5 to 50%	316	105	32	11
50 to 95%	518	175	52	17
95 to 99.5%	1100	367	110	37
0.5 to 99.5%	2171	724	217	72
(b) an allele of recessive phenotype				
Selection	1%	3%	10%	30%
Frequency change				
0.5 to 5%	1100	367	110	37
5 to 50%	518	175	52	17
50 to 95%	316	105	32	11
95 to 99.5%	237	79	24	8
0.5 to 99.5%	2171	724	217	72

Calculated from Table II of Haldane (1924).

Even in the well known case of the peppered moth (*Biston betularia*) where the selective force against the recessive is of the order 0.2, a 20% selective disadvantage, the change from low frequency of the dominant (possibly introduced by mutation, possibly pre-existing in the population at a very low frequency from a polymorphism), took over 50 years, that is to say over 50 generations (Kettlewell 1958). All this refers only to individual selection. Group selection, where one whole population is selected at the expense of another will not be discussed. This topic has been very popular, and even MacArthur and Wilson (1967) have urged that it might apply to island populations. My reason for ignoring it here, is that it has been very fully discussed by Williams (1966), and I find that his arguments that group selection will scarcely ever be of any importance, even in the type of situation envisaged by MacArthur and Wilson to be convincing. Some cases where individual selection produces results that appear to a superficial glance to be due to group selection, have been discussed by Hamilton (1964).

Gene frequency will also be a subject to random effects of the type called drift by Sewell Wright and also to other sorts classified by Wright (1955). The discussion of these effects has played a large part in genetical theory, while the corresponding effects have been largely ignored in ecological theory. It seems that the difference is not simply an historical one, but that random effects might well be more important in genetical systems, simply because of the neutrality of the Hardy-Weinberg equilibrium. If a population is in Hardy-Weinberg equilibrium and some random happening changes it to a new gene frequency, it will come to a new p^2 to $2pq$ to q^2 equilibrium at that gene frequency; there will be no tendency to return to the old one. In contrast equilibrium population sizes are normally stable, and any displacement will be compensated. The chances of a random extinction of a population under the logistic curve have been studied particularly by Bartlett (1957). Leslie (1958) gives a numerical example in which a population of Paramecium of only 100 individuals at equilibrium would be expected to become extinct from random fluctuations of population size in approximately one million years. That is to say when the population system involves a stable equilibrium as in the logistic theory, rather than a neutral equilibrium as in the Hardy-Weinberg system, the chances of random changes being of any importance are totally negligible. Polymorphic equilibria are also of the stable type, so again for a first analysis there is little point in considering random effects. That is, I agree with Moran (1967) that the importance of random effects in genetic systems have been much overstressed.

Balanced genetic systems

The clear demonstration that most individuals of a population are genetically distinct has only really followed the extensive use of gell electophoresis (Manwell and Baker 1970). Situations in which there are two alleles in a population, or two other genetic forms, are called polymorphisms, and where the coexistance results from a balance of forces, not necessarily selective forces, then there is a balanced polymorphism. The importance of the topic was first strongly urged by Ford, and to emphasise this, I must quote his classical definition (Ford 1945) 'that polymorphism is the occurence together in the same habitat of two or more distinct forms of a species in such proportions that the rarest of them cannot be maintained merely by recurrent mutation.' Ford has pointed out cogently and correctly that the study of polymorphism gives the best opportunity for the study of selection in natural populations. The classical studies on the mathematical theory of population genetics by Haldane, Fisher and Wright, all assumed a selective force of .01 or less. Studies on polymorphism have shown that the selection on major genes occuring in populations can be many times larger than this, as in the transient polymorphism of *Biston betularia*. There are many other cases where selective forces are normally over 0.1, that is to say, 10 times as large as those considered in the classical papers (Ford 1964).

Consider first the simplest, most wide-spread example of polymorphism, the dimorphism of male and female. The commonest genetic basis of this is that females carry two X chromosomes (XX) while the males are XY. In most species the nature of the organism ensures that the male must mate with the female and vice versa, and the male produces equal numbers of X and Y gametes. Because of this, irrespective of the proportion of males and females in one generation, in the next there must be a 1:1 ratio of XX to XY genotypes, or in other words equal numbers of males and females. It is not possible to produce homozygote YY's. Here the balance is produced by segregation and the mating system. This example illustrates too that there are underlying biological problems quite different from each other in all problems in population analysis. The reason for segregation of X and Y gametes, and the reasons for having sex at all (Williams 1966, Maynard Smith 1968), are problems involving a much longer time scale, and so involving a different type of answer (Haldane 1948). The first of these problems is a developmental one, eventually leading back to questions in molecular biology (Lewontin 1968a), but the second one is merely a deeper problem in population genetics.

The problem of studying the change of gene frequency resulting from selection is most easily approached by studying the change in gene frequencies. There is an analogy with the study of the change of population number which was considered in Chapter 5. For the sexual mating system, as the system comes back to an equilibrium of 0.75 in every generation the graph of Δq (the change of gene frequency) against q (the gene frequency) consists of a straight line with a slope of -1, i.e. running at 45°, and intersecting the q axis at 0.75. With only two genotypes, one could equally construct a figure for change of genotype frequency. In the slightly more general case of three genotypes for two alleles, an extra dimension is needed. This can be done by using the de Finetti diagram, which is a triangular graph with the three axes corresponding to the frequency of the three genes, and the use of this has been shown skilfully by Cannings and Edwards (1968).

Three curves of Δq against q, all relating to the production of a balanced polymorphism with a frequency of one allele of 0.3 and the other of about 0.7 are shown in Fig. 6.1. These three systems relate to three different experimental systems using beads or other counters to imitate the behaviour of a population. The approach to the equilibrium can be studied in the same way as the Δn curves on p. 61. As the Δq axis has been expanded relative to the q axis for clarity, all three curves have a maximum slope appreciably less than -1 and lead to a smooth approach to the equilibrium state from any position. They illustrate

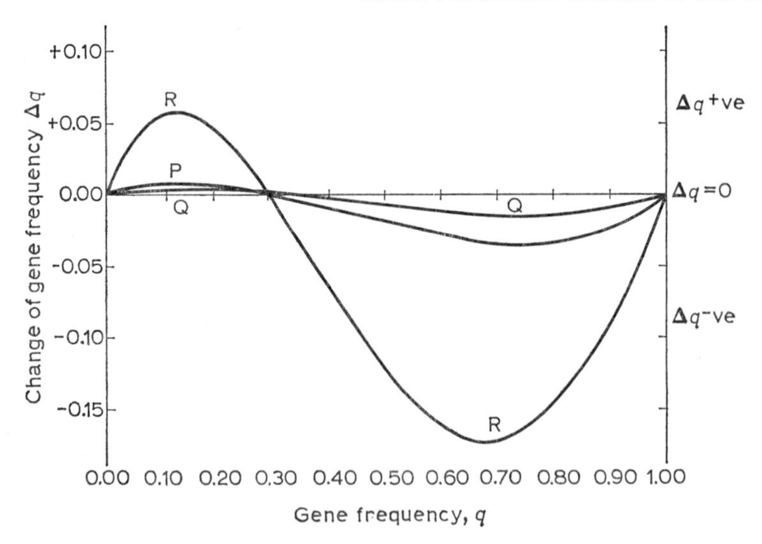

Fig. 6.1 Curves of change in gene frequency (Δq) against gene frequency (q) for three balanced polymorphisms. Note the differences in the scales along the Δq and q axes. The equations of the curves are given in Table 6.2. For further explanation see the text.

three rather different ways of maintaining a polymorphism. The curve P is the classical system of heterozygote advantage, apparently first formulated by Fisher (1931), in which each homozygote is at a selective disadvantage. In this particular case the selective disadvantage of one genotype is 0.1 and of the other 0.25. System Q can be described as a secondary polymorphism because it involves a sex-linked allele. The females are diploid with three genotypes a_1a_1, a_1a_2, a_2a_2 and selection of 0.5 is applied against a_1a_1 and 0.1 against a_1a_2. The males are hemizygous with the two genotypes a_1 and a_2 and selection is 0.2 against a_2. That is to say, selection is acting differentially in the two sexes. The third system, R, depends on mating preference. Again there are diploid females and hemizygous males. In this experiment the beads are counted off in fours. The first 2 represent a female and the second two two males from which she will choose a mate. Even if the two males are of the same genotypes she only chooses one of them. If the two males are different she chooses the one of opposite phenotype, and we assume that the allele a_1 is completely dominant, so that the phenotype a_1a_2 in the female is the same as the phenotype a_1a_1. So both a_1a_1 and a_1a_2 choose an a_2 male if they can, while the a_2a_2 females choose a_1 males. The maximum selection in such a system is clearly 0.5, and the selection should normally be less, of the order of 0.25.

There are some other interesting points about the three curves *P*, *Q* and *R*. As their selective systems are different, there is no simple way of comparing the selective forces, but they are all more or less of the same order. Nevertheless the curves of Δq differ markedly in amplitude. That is, the rate of change of gene frequency will depend not only on the force of selection but in the way in which it is applied. Considering the natural mechanical analogy with the application of forces to levers, this result is not surprising, but seems not to have been stated before. Another point is that the curves all appear to be cubical equations, which have three points where they intersect the axis of $\Delta q=0$. The two roots at $q=0$ and $q=1$ are unstable in the quite simple sense that if the gene frequency is slightly

Table 6.2

System P	$\Delta q=\frac{2q-9q^2+7q^3}{18+4q-q^2}=\frac{q(1-q)(2-7q)}{18+4q-q^2}$
System Q	$\Delta q=\frac{q-4q^2+3q^3}{18-3q^2}=\frac{q(1-q)(1-3q)}{18-3q^2}$
System R	$\Delta q=q-5q^2+6q^3-2q^4=q(1-q)(1-4q+2q^2)$

inside these values it will change progressively to reach the equilibrium. The equations for the curves are set out in Table 6.2, where it can be seen that none of the equations is a simple cubic, all are slightly more complicated. The equation for Δq against q normally has a root at $q=0$ and $q=1$, and so the function for Δq can be made a simpler one by dividing it by $q(1-q)$. This is discussed by Li (1955a) and below in Chapter 7 in relation to Panaxia. This is because homozygous populations are normally viable, and cannot be seeded with another allele by selection, but only by mutation or migration. An exception is found in the self-sterility alleles of some plants. When discussing Δn against n curves, it was shown that if the slope of the curve was greater than -1 but less than -2 the equilibrium would be reached by damped oscillations. The same considerations apply to Δq curves (Lewontin 1958, Cannings 1969) and can be applied to cases of self-sterility alleles in plants. In these systems there are a number of alleles, s_1, s_2, s_3 etc., and a gamete carrying an allele s_i can only fertilize a diploid individual s_js_k where s_j and s_k are any alleles other than s_i. This means the system must have a minimum of three alleles, and that all individuals are heterozygotes. Homozygous populations cannot exist. The equation for one allele in a three allele system is (Kempthorne 1957)

$$\Delta q=\tfrac{1}{2}(1-3q)$$

which is a straight line of slope—1.5 and so the frequency of an allele starting below the equilibrium frequency overshoots by 50%. If the deviation is $-x$ to start with, the next one is $+\frac{1}{2}x$, the third one $-\frac{1}{4}x$, and so on, the equilibrium being reached by a damped oscillation.

One final point of theory can be considered before discussing, in the next Chapter, some observed polymorphisms. This is, is it possible to classify the systems that can produce a balance polymorphism? Some attempts at classifying this have been made by Haldane (1954) and Williamson (1958, 1960). None of these attempts is fully satisfactory, because they seem to use arbitrary categories, and give no promise to be all inclusive. A satisfactory classification would indicate the limits of the possible systems that might exist for producing polymorphisms, as well as listing systems that have been suggested, and would show why other systems could not be effective. Both Haldane's and my systems were verbal and it is possible a more effective method might be found through using functions and equations of the type shown in Table 6.2. This had not yet been attempted. Probably neither of these approaches is likely to be effective, and that there are more fundamental problems to be settled. Consider the classic and most discussed mechanism for polymorphism, that produced by selection for the heterozygote. In some cases one might actually select for a heterozygote. The simplest case is a breeder trying to produce say a pure bred line of some form found only in heterozygotes. Genetic segregation will ensure that he fails. On the other hand a case often referred to as heterozygote advantage is biologically distinct. This is the polymorphism of sickle cell in human populations in malarious regions of Africa and elsewhere. Typically there are two alleles Hb^A and Hb^S. Hb^A Hb^A individuals contain normal haemoglobin. In Hb^S Hb^S the molecular confirmation of haemoglobin changes so that the erythrocytes become sickle shaped, and individuals normally die of anaemia. The heterozygotes are said to show the sickle cell trait. Heterozygotes are less susceptible to falciparum malaria than homozygotes. Heterozygotes are at an advantage to both homozygotes and this produces a polymorphism and the selective forces seem to agree with the equilibria observed (Allison 1964). The important point about this system is that it is not a simple advantage of the heterozygote but two biologically distinct disadvantages for the two homozygotes. Mathematically the systems are the same, but the biological distinction seems worth making. They could perhaps be distinguished as heterozygote advantage and dihomozygote disadvantage. This suggests a classification as follows:

1 *One selective force: balance produced by*

(a) segregation—heterozygote advantage
(b) variation in the strength of the selective force depending on frequency—apostasy and similar systems (Clarke 1962, 1969)
(c) selective force varying with the density of the population, density dependent selection (Turner and Williamson 1968).

2 *Mating systems*

(a) sex
(b) other out-breeding systems
(c) self-sterility alleles.

3 *Two selective forces: balance determined by*

- (a) genetical systems
  - (i) dihomozygote disadvantage
  - (ii) sex linkage
  - (iii) different selective values in the different sexes
- (b) two selections of different functional form (Williamson 1960)
  - (i) mixture of mating preferences and viabilities
  - (ii) mixture of mating preferences and fertilities
  - (iii) mixture of frequency or density dependent selection with constant selection in the opposite direction
  - (iv) mixtures of frequency and/or density dependent selection of different functional form

Note that heterozygote advantage and dihomozygote disadvantage are only two of many possibilities that exist for maintaining a polymorphism of two alleles at one locus.

The place of any real polymorphism in the classification can be known if the biological forces producing it have been established. Some indication of the place could be got from knowing the rate of change of gene frequency from a given gene frequency and also the intensity of selection at that gene frequency, but as will be seen in the next Chapter, these are usually unknown.

7 Observed polymorphisms

The analysis of natural genetical systems is easiest when the genes are major genes, in the sense of Rendel (1968). Then there is an exact relationship between phenotype and genotype, and normally there is strong selection. This situation is that typically found in polymorphisms. These have been extensively studied both by Ford, Sheppard and their collaborators in England and by Dobzhansky's school in America. All those cases in which an analysis is more or less complete have been mentioned in Chapter 6. These fall into four sets; sex and other outbreeding systems (Finney 1952); self-sterility alleles; the sickle cell system in the haemoglobin of man; and the transient polymorphism in industrial melanism in the peppered moth *Biston betularia*. The major features of all these systems seem to be known, and the observed equilibria seem to agree fairly well with the selective forces measured. There are inevitably still some problems. In Biston the status of the insularia melanic form is not clear, nor is it clear whether even in industrial areas a balanced polymorphism has not been produced, or whether the transitory polymorphism has got stuck in a slow phase (Kettlewell 1958).

This Chapter will be concerned with four other cases, two from America and two from Europe, which have been studied very extensively, and which are still far from fully understood. It is indeed because of this that they each present interesting features about the problems to be solved in analysing the genetic variability in natural populations. The four systems are the t-locus in the house mouse, *Mus musculus* studied by Dunn and his associates in New York; the medionigra polymorphism in the moth *Panaxia dominula*, studied by Ford, Sheppard and their associates from Oxford; the chromosomal polymorphism in *Drosophila pseudoobscura*, studies by Dobzhansky and his associates working from New York; and fourthly the polymorphism in the sibling species of snails *Cepaea nemoralis* and *Cepaea hortensis*. The last system has attracted biologists since the 19th century but our present knowledge comes mostly from the work of Cain and Sheppard and their associates in England, and Lamotte and his associates in Paris. In all four cases stable equilibria are observed, but in none of the cases are the forces that have been measured sufficient to explain the values of the equilibria. Only in Panaxia is there the simple situation of two alleles at one locus with the heterozygote distinguishable from both homozygotes.

t alleles in *Mus*

Of these four systems, that of the tail polymorphism in the mice seems the most simple, and the most likely to produce results that will match the theory. It is rather curious in that its detection and study depends entirely on the existence of a genetic variant that has only ever been found in the laboratory, the dominant T for brachy. The homozygous wild type mouse ++ has a normal length tail, the heterozygote with brachy +T has a short tail, as the name implies, about three-quarters of the length of the ordinary tail, while the homozygote TT is lethal. In many wild populations of house mouse, or rather in non-laboratory populations as they usually occur in and around houses, a series of other 'alleles' at this 'locus' is found. These are all recessive and designated collectively as *t*, with various superscripts to indicate the different types of *t*. An important point is that the heterozygote +*t* has a normal phenotype; it is apparently not phenotypically distinguishable from ++. The homozygote *tt* is frequently lethal, but sometimes only sterile, while the detection of the system depends on the fact that the other heterozygote *Tt* is a tail-less mouse. Unfortunately mouse chromosomes are difficult to study. The fine structure cannot usefully be studied under the microscope, nor is it feasible to rear so many mice that rare cross-over events can be recorded accurately. Nevertheless, Dunn's studies have involved a very large number of mice, and no cross-overs have been recorded. In view of this, and the large number of *t* 'alleles' that have been detected, it seems likely that the *t* phenomenon involves some major rearrangement of a section of the chromosome, such as a series of inversions, rather than the simple mutations at one cistron.

A large number of different *t* alleles has been recovered from wild populations of house mice, and these are designated t^{w1}, t^{w2} etc. The different alleles are distinguished both by their different pattern of lethal effects on homozygotes and by their different effects when combined as heterozygotes. The developmental consequences of a different *t* allele is an important topic which may well lead to a better understanding of the selective forces, but has not yet reached the stage where it helps the analysis, and so unfortunately cannot be discussed here. It must suffice to say that the homozygote $t^w t^w$ typically causes the embryo to die after a few days of uterine life. As the homozygote $t^w t^w$ is lethal, there is tautologously, strong selection against it.

Only one other important selective force has been discovered. This is that in the heterozygote males, +*t*, there is an abnormality in segregation. Instead of producing 50% offspring carrying + and 50% carrying *t*, up to 90–99% of their offspring carry *t*. There is no abnormality of segregation in the heterozygote females, and the segregation differential in the males apparently comes neither from meiotic drive (an abnormality of the meiotic system producing an excess of one gamete or another), nor from differential mortality in utero of the embryos once they have been formed. It seems to be straightforwardly a case of differential transmission, that is to say the differential success of different sperm in fertilizing ova. There has in fact been an interesting experiment recorded by Dunn (1960) by Braden. Normally, mating occurs five hours before ovulation and eight hours before fertilization. If mating is restricted to a period after ovulation, then the segregation ratios are changed, and changed towards 50%, though they do not in fact reach it.

So here is a case with one selective force acting one way on the diploid part of the life cycle, and a different selective force acting on the haploid part. The deterministic mathematics of this system have been worked out by Bruck (1957). If m is the proportion of effective t's in the sperm, and $\hat{p}$ is the equilibrium frequency of t type alleles in adults, then

$$\hat{p} = \frac{1}{2} - \frac{(m(1-m))^{\frac{1}{2}}}{2m}$$

Bruck showed that this was a stable equilibrium and, as there are no other stable equilibria with these two forces of selection, that the system would be expected to go to this equilibrium from any gene frequency. That is, one t allele introduced into a pure wild type population, or the biologically almost impossible one wild type allele introduced to a pure t population, would both lead to the population ending up at the stable equilibrium. It is easy to see that if $m=0.5$, i.e. if there is no segregation advantage for the t allele, then the equilibrium frequency of t is zero, as one would expect. Any segregational advantage leads to some equilibrium. A total segregational advantage, with $m=1$, so that all the offspring of the $+t$ male heterozygotes carrying the t allele and none the $+$allele, leads to an equilibrium of only $p=0.5$. The segregational advantage of .9 leads to an equilibrium of .333 ($=1/3$), and most t alleles found in wild populations have a segregational advantage which is greater than this.

The actual frequency of t alleles in wild populations is lower than one would expect from this theory. The observed frequency is usually in the range 0.25–0.35. Clearly some other factor must be brought in. This might be a straight forward selective factor (see Lewontin, 1968b and Young 1967). There is unlikely to be any simple form of heterozygote advantage, because that would lead to a higher frequency of t rather than to a lower one, but these could of course be a heterozygote disadvantage, or an increased advantage to the wild type homozygote. As the discrepancy is not exceptionally large, the effects causing it may not be large, and if there are several such effects all adding together to produce the discrepancy, each of them may be individually rather difficult to trace. The fact that no phenotypic difference is known between the $+t$ and $++$ mice would suggest this is so; but against this, the fact that there is an appreciable discrepancy suggests that an explanation ought not to be too hard to come by.

Lewontin and Dunn (1960) have suggested another type of explanation. Working from the well known fact that house mice occur in fairly small groups in any one generation, they have postulated that these small populations are persistent, and being so small will be subject to random sampling effects as well as to selective effects. In these small populations, any population in which t goes to fixation will automatically become extinct, while one in which the wild type allele goes to fixation will persist, and remain homozygous until a new t allele is introduced by migration or mutation. So using a somewhat improbable persistent population of 6 females and 2 males, a computer simulation showed that there would be a U-shaped distribution of gene frequencies in populations, with one peak at $p=0$ and another at p between .4 and .5. The two peaks correspond to the two equilibria, that between 0.4 and 0.5 being the stable equilibrium, and that at around 0 being the unstable equilibrium reached by the random extinction of t alleles. This explanation suffers from two disadvantages. The first is the assumed persistent small and unbalanced population. The actual distribution reached is sensitive to the population size, and so this explanation is not in that sense a robust one. There is an assumption that migration rates are rather small but this is contrary to some English experience of house mouse populations, which in summer become much more mixed, and does not allow for the inevitable migration of the young mice (Southern 1954). The second drawback is that it is assumed that the observed frequency of .25–.35 in wild populations comes from averaging a number of populations, and is not the true frequency in most of them. With the difficulties of scoring t alleles, involving laboratory stocks carrying T, this suggestion is plausible but open to testing. In a later paper, Dunn and Bennett (1967) produce evidence that suggests it is in fact improbable.

This evidence came from a laboratory 'wild' population started in 1944 with New York City, Long Island and Philadelphia mice. In the early years this population contained +, t^{w1} and t^{w2}. From 1944–1950 the population size was kept at between 75 and 125, while from 1951–1959 at a rather smaller size between 50 and 60, but nevertheless very much larger than the sizes considered by Lewontin and Dunn. The condition of the population allowed for two generations per year. t^{w1} was lost sometime between 1952 and 1955. The transmission ratios t^{w2} in 1952 was 0.954, and this may conceivably explain the extinction of t^{w1}, if t^{w2} was at an appreciable selective advantage to it. However, the transmission ratio t^{w2} measured in 1955 and 1959 was only 0.84, while the frequency of t in the populations during those years was 0.37. With the lower transmission ratio the expected frequency of t is only 0.282 and so t^{w2} was at a higher fre-frequency in this population as might be expected. This is in contrast to the situation which Lewontin and Dunn are trying to explain where t alleles appeared to be at a lower frequency than Bruck's theory predicted. Bruck's theory would give for the transmission ratio observed in 1952 an equilibrium gene frequency of 0.39, only slightly higher than the frequency in 1955 and 1959. As far as this single population goes, it suggests that an equilibrium frequency between .25 and .35 may in fact be a perfectly genuine one, and that there may be variation in transmission frequency. The work mentioned earlier on the variation with transmission rate with time of fertilization suggests that this might be an important factor in natural populations, leading to a lower equilibrium frequency of the t allele than would be expected from the measurement of the transmission ratio in laboratory conditions. The question of what really happens to wild populations must be studied not only by more detailed work of the structure of wild populations and the amount of interchange between them, but also be a search for other selective factors, and by observations on the mating habits of wild mice, because this may affect the real transmission ratios.

Panaxia dominula

About half the populations of wild mice studied by Lewontin and Dunn, 14 out of 38, were polymorphic for a t^w allele. Nearly all populations of *Drosophila pseudoobscura* (Dobzhansky 1951) and of *Cepaea nemoralis* (Lamotte 1951) are polymorphic, but the next case is quite unlike these. *Panaxia dominula*, the scarlet tiger moth, is found in discrete colonies scattered over the southern half of the British Isles (Kettlewell 1942, Cook 1959), and elsewhere in Europe.

In only one of these colonies, that at Cothill, near Oxford, is there a natural polymorphism at the medionigra gene. The unusual allele at this locus produces a fairly small effect in the heterozygotes, which can normally be scored quite easily and has, in the manner of the Lepidopterists, been given the latin name *medionigra* while the homozygote for this allele produces a much more striking form, the so called *bimacula*. This population has been studied ever since 1939 by Ford, and much of the work has been done by Sheppard. Their latest paper (Ford and Sheppard 1969) and Lees (1970) gives the population figures that enable Fig. 1.14 (page 12) to be nearly up to date. The gene frequency of medionigra at Cothill has been below 5% since 1947, and so the frequency of bimacula is too low to come into consideration: only a dimorphism of the wild type and the medionigra heterozygote need be considered.

Sheppard (1952, 1953) found four selective forces. One was a mating preference leading to wild types mating with medionigra more frequently than could be expected. The second was a lowering of fertility of medionigra males. The third was a higher larval mortality in medionigra, though this was measured not at Cothill but at a neighbouring artificial colony. The fourth was bird predation acting against bimacula. Only the first of these, as Sheppard has pointed out, can serve to increase the gene frequency of medionigra; the other three all act against it, even though the last may be neglected. The mating preference is difficult to measure, and indeed is not clear what is the best experimental way of measuring it. Sheppard and Cook (1962) state that larval mortality would be expected to vary appreciably with conditions. They give results for the frequency of medionigra gene not only at Cothill but in a number of artificial colonies. These are shown in Fig. 7.1 where $\Delta q / q\,(1-q)$ is plotted against q. This is the form suggested by Li (1955a) and as was pointed out in Chapter 6, can be made statistically respectable by plotting q_{i+1} against q_i, even though the data do not stand out so clearly in that form. It is not the weakness of the particular form of the data chosen for plotting in Fig. 7.1 but the rather small number of results inevitably available at each colony which prevents one saying that there have been a number of distinct equilibria in different places, or even, at Cothill, in different periods of years. Should the equilibrium be essentially one between mating preference and larval mortality, then a stable polymorphism could be produced (Williamson 1960) though the actual equilibrium frequency at any one place at any one time has still not been explained quantitatively. The reason for the polymorphism being confined to one particular population then would be that

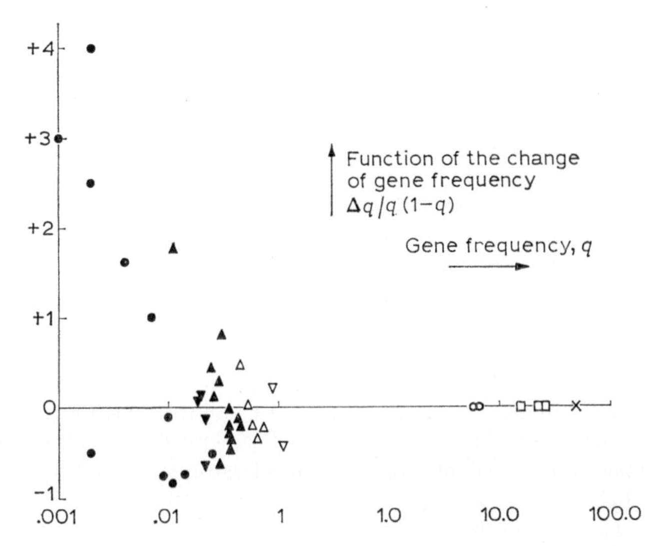

Fig. 7.1 Scatter diagram for the change of gene frequency (Δq) against gene frequency (q) for populations of *Panaxia dominula* at different places and different times. For clarity, the x axis is log q rather than q, and the y axis is the function $\Delta q/q(1\text{-}q)$. Each set of points distinguished by a different symbol appears to be spread about a different zone of the x axis. The observations on gene frequencies are indicated as follows

●	Sheepstead Hurst, 1954–1968
▼	Cothill 1959–1963
▲	Cothill 1946–1958
△	Cothill 1941–1945
▽	Cothill 1939–1940
○	Hinksey 1959–1961
□	Genetic garden 1958–1961
X	Ness 1960

Data from Ford and Sheppard (1969), Sheppard and Cook (1962)

the differential in larval mortality is only small enough in that population for the polymorphism to persist. Elsewhere, one might presume, that should the medionigra gene be introduced by mutation or, rather improbably, by migration from the Cothill colony, it would have been eliminated quickly. However, medionigra has been introduced deliberately into the nearby colony of Sheepstead Hurst (Ford and Sheppard 1969) and it has persisted there at a very low frequency, and the data on the colony are also given in Fig. 7.1.

Drosophilia pseudoobscura

The chromosomal polymorphism in *Drosophila pseudoobscura* has been described in great detail by Dobzhansky (1951) and is described in outline in almost any book that deals with evolution. There are at least 19 overlapping inversions on the third chromosome known in *Drosophila pseudoobscura* populations, and one, a hypothetical type corresponding to that found in the related species *D.miranda*, which can be presumed to have existed in *D.pseudoobscura* at one time. When they were first discovered in the late 1930's, the occurrence and variation in frequency in these inversions in different populations was ascribed largely to genetic drift. Mayr (1945) seems to have been one of the first to point out that selection must be important in this species, and this is now accepted as being a major factor in determining the frequencies of different inversions. There are four striking phenomena which are consistent with a selective explanation. The first is that the different inversions have characteristic geographical distributions. The second is that the frequencies of the inversions vary systematically with altitude. The third is that at any one place there is frequently found a regular clear cut annual cycle in the frequencies, and the fourth is that there has been, in the south west U.S.A., a systematic change in which the inversion Pikes Peak has replaced the inversion Chiricahua to a large extent. (Dobzhansky 1958).

With so many inversions to consider, it is scarcely surprising that no full mathematical analysis of this polymorphism has been attempted. Still, a stronger reason for this is that remarkably little is still known about the ecology of Drosophila. Despite many laboratory experiments, the selective advantages that have been shown to be associated with different inversions have not been analysed biologically. This task is difficult, because the inversions themselves evolve. The clearest proof of this is that the same inversion collected in different areas behaves differently in laboratory populations. So there are not only a lot of genetic variables, but all of these themselves vary, and the important action of any one of them is still not known.

Gene changes consistent with heterosis have been shown to occur frequently. But heterosis cannot be the whole cause of the maintenance of this polymorphism. The first argument against it is that some inversions occur at low frequencies. In the case of heterozygote advantage, the slope for Δq against q at the equilibrium point is—$st/(s-st+t)$, where the selective advantages of the three genotypes a_1a_1, a_1a_2, a_2a_2 are $1-s$, 1 and $1-t$ (Williamson 1960). The equilibrium gene frequency is $s/(s+t)$ and consequently for rare genes, and when t approaches 1, Δq and $\hat{q}$ are both nearly equal to s. That is to say if the equilibrium is 1% the slope of Δq against q is less than 1 in 100, and this is barely stable. So polymorphisms maintained at low frequency probably involve some forces more effective at these frequencies than heterozygote advantage. I will return to this point when considering Cepaea.

Another reason for supposing that heterozygote advantage is not the sole explanation comes from triple experiments with three inversions in the laboratory (Dobzhansky 1957). Experiments were done with the three inversions ST (Standard) AR (Arrowhead) and CH (Chiricahua). The selective values estimated from populations with pairs of these inversions and that from triple inversions are shown in Table 7.1. Taken in pairs, all combinations are heterotic,

Table 7.1 Selective values in experimental populations of *Drosophila pseudoobscura*

(a) with two inversions only

ST/AR	1.0	ST/ST	0.8	AR/AR	0.5
ST/CH	1.0	ST/ST	0.8	CH/CH	0.5
AR/CH	1.0	AR/AR	0.9	CH/CH	0.5

(b) with all three inversions

ST/ST	ST/CH	ST/AR	CH/CH	AR/CH	AR/AR
0.83	0.77	1.00	0.36	0.62	0.15

(Data from Dobzhansky 1957)

ST appears to have the same advantage over AR and CH, but in the AR CH combination AR is at an appreciable advantage to CH. In the triple state the situation is rather different. ST and AR considered as a pair are still heterotic, but the AR AR homozygote is at a very much greater selective disadvantage than it was before. ST with CH is no longer heterotic at all, and the AR CH combination, while still heterotic, now gives an advantage to the CH CH homozygote. The selective values are population dependent, and is not reasonable to suppose that even in the two inversion experiments they are constant.

The population of *Drosophila pseudoobscura* for which there has been the most thorough analysis of selection is a very isolated one in Colombia (Dobzhansky *et al.* 1963). This population has the considerable advantage for an analysis of having only two inversions ST and TL (Treeline). Two forms of selection were detected from observations of the frequencies of laboratory broods. The first could be a straight forward heterozygote advantage. From broods that should be segregating 1:1 or 1:2:1 the frequencies are given in Table 7.2. In all cases there

Table 7.2 Observed numbers segregating in offspring of the Colombian population of *Drosophila pseudoobscura*

Segregation	Expected frequencies	Observed numbers		
ST/TL : SC/SC	1 : 1	590		463
ST/TL : TL/TL	1 : 1	168		112
ST/TL : SC/SC : TL/TL	2 : 1 : 1	165	77	66

(Data from Dobzhansky et al 1963)

is a surplus of the heterozygotes, and the selective values appear to be ST/TL 1 ST/ST 0.8 TL/TL 0.7. This is of course quite sufficient to produce a polymorphism by itself. From wild females and the progeny they produced it is possible to classify the frequency of different types of mating. These frequencies can be compared with those expected, since the chromosome frequencies of the two forms in the wild are known to be ST 55.6 and TL 44.4. The actual frequencies observed are shown against those expected in Table 7.3. It can be seen

Table 7.3 Mating frequencies of wild females of *Drosophila pseudoobscura* in the Colombian population, judged by the segregation of their offspring

	Matings with homozygotes only			Matings involving heterozygotes		
	ST/ST × ST/ST	TL/TL × TL/TL	ST/ST × TL/TL	ST/TL × ST/ST	ST/TL × TL/TL	ST/TL × ST/TL
Observed	42	4	31	159	36	47
Expected	59.0	4.6	33.2	123.5	34.6	64.6
% change of observed from expected	−29	−13	−7	+29	+4	−27

(Data from Dobzhansky et al 1963)

that all matings in which one genotype mates with a male of the same genotype, whether those are homozygotes or heterozygotes, occur with lower frequency than expected, and the same applies to the joint mating with two homozygotes. The only matings in which there is an excess of observation over expected are those where one homozygote mates with the heterozygote. These data are not corrected for the segregation that might be overlooked either in small broods or as a result differential viability, but these would both tend to increase the number of broods scored as not segregating, and these are all those in which there are deficiencies. So, rather as in Panaxia, there appears to be a strong tendency for the heterozygote and homozygote to mate with each other rather than to mate at random. This type of mating, as in Panaxia, will come out mathematically as a variable selective force and so could conceivably explain the experimental results with three inversions simultaneously. There is at least no lack of selective forces that should produce equilibrium in Drosophila. There is, however, a lack of a quantitative model which could examine the forces to see if they fit the observed data.

There is another set of experiments which suggests that there are in fact other selective forces also involved in the polymorphism in Drosophila, experiments on the 'fitness of populations'. Fitness, and adaptation, are two concepts that are defined in relation to individuals, and which have no clear meaning when applying to populations (Cain and Sheppard 1954, Li 1955b, Williams 1966); and the various measures of the population which might be thought to be reasonable extensions of the concept of fitness of an individual, do not in fact correlate well with themselves (Barker 1967). Nevertheless, starting with the enigmatic statement by Dobzhansky (1951) that polymorphism increases the efficiency of the exploitation of the resources of the environment by the living matter, Dobzhansky, *et al.* (1964) showed that polymorphic populations of *Drosophila pseudoob-*

scura are more successful in a variety of ways than monomorphic populations. They tend to have a greater rate of increase, and tend to produce a greater equilibrium population size. These experiments all suggest that the different genotypes are, to a reasonable extent, acting as different independent populations. If this is so, then there must be density dependent selection. Ohba's (1967) experiments tend to confirm this, as the differences between the populations almost disappear under 'near-optimal' conditions, that is, with effectively unlimited growth and an absence of density dependent interactions. Laboratory populations of *Drosophila pseudoobscura* seem to provide the most promising data for trying to fit the change of gene number into the fundamental equation of population dynamics $\Delta n = b - d$.

None of the three polymorphisms discussed so far in this Chapter are particularly striking to the casual observer. The *t* alleles in mice are not yet recognizable by any phenotypic character but have to be scored by breeding. Chromosomal inversions in Drosophila require the rearing and squashing of larvae, while in the polymorphism of Panaxia there is a moderately small difference between the medionigra heterozygote and the wild type dominula, and this is confined to one population. Compared with industrial melanism or sex, they do not leap to the eye, and it might be thought that this difficulty in detecting them might be one reason why they are not yet fully understood. The final example in the Chapter contradicts that suggestion.

Cepaea nemoralis

The polymorphism in shell colour and pattern in Cepaea is a most striking one and any one who has ever seen a collection of Cepaea shells can appreciate at once the attraction of this system for study. Several species of European snails and slugs are polymorphic for their external characters, but the two species of Cepaea have much the most striking polymorphism. *Cepaea nemoralis* has been studied more than *Cepaea hortensis*, but the polymorphism shows a remarkable parallelism in the two species. The basic shell colour may be various shades of yellow, various shades of pink grading into quite deep reds, or light or dark brown. On top of this there may be a banding pattern. The typical form has five clear cut bands, which are numbered from the top downwards 1–5. Almost any one or more of these may either be missing, or fuse with its neighbours, producing a wide variety of shells. In *C. nemoralis* the forms with no bands, 00000, the form with the one central band, 00300, and the form with the three underbands, 00345, are the commonest variants. Fusions of bands occur often between 1 and 2 and between 4 and 5, and sometimes between 3, 4 and 5, and the whole set of five bands may be fused to form one, largely brown, very broad band. The scoring of fusions is difficult and arbitrary, as the fusion may start at any point down the shell, but using a standard convention, Cain and Sheppard (1950) give the frequency of individual shell types in a number of colonies. Most authors since then have not attempted complete tabulations, as these seem not to be particularly informative. The loci that have been analysed genetically are shown in Table 7.4. The complex locus for shell colour (C,) and the other important

Table 7.4 Genetic loci known to be involved in the polymorphism of the appearance of the shell of the snail *Cepaea nemoralis*

Locus	Feature	Alleles, listed from dominant to recessive, and each dominant to the alleles to the right, and phenotype of homozygote.
C	Shell ground colour	C^B brown, C^{DP} dark pink, C^{PP} pale pink, C^{FP} faint pink, C^{DY} dark yellow, C^{PY} pale yellow. Also possibly pale brown, faint brown and yellow-white.
B	Banding	B^0 no bands, B^B banded (i.e. one or more depending on other loci). Closely linked to C.
U	Unifasciate	U^3 00300, U^- more bands.
T	Trifasciate	T^{345} 00345, T^- bands 1 and 2 as well.
S	Spread bands	S^S spread bands, S^- normal. Closely linked to B and C.
I	Interrupted bands	I^I punctata (bandes pâles of Lamotte 1951), I^- normal. Possibly closely linked to B and C, see Cook (1967).
P	Pale bands and lip	P^N normal dark bands and lip, P^L slightly pigmented bands and lip, P^A white lip normal bands, P^T white lip, transparent bands, shell colour pale (hylozonate). 10–15% crossover with B and C apparently. (Cook 1967).
O	Orange bands	O^- normal, O^0 orange bands (lip normal).
R	Darkening bands	R^- normal, R^D darkening down the shell.

Mostly from Cain, Sheppard and King (1968). See also Cook (1967) and Wolda (1969b).

locus for presence or absence of bands, are closely linked, and those who like the term can refer to the pair as a supergene. The rather rarer S locus, is also very closely linked. The loci labelled I and P are linked, but not so closely, and cannot easily be called part of the supergene, while all the other loci, including the important dimorphism for the one and the three banded forms, U and T, are not linked. As well as the forms listed in Table 7.4 it is known that fusion of bands and size of shells are controlled genetically, though segregations have not been scored.

Cepaea nemoralis is hermaphrodite and a perennial. It reproduces by laying clutches of eggs which are buried in the ground. The young snails seem to spend their early life underground rather than crawling on the surface, and only half grown and larger individuals are normally included in population counts. The best data illustrating the life history and the discussion of the time required to come to maturity are given by Wolda (1963).

Almost all populations of Cepaea are polymorphic, though Lamotte (1951) noted one or two rather small colonies that are apparently monomorphic. The

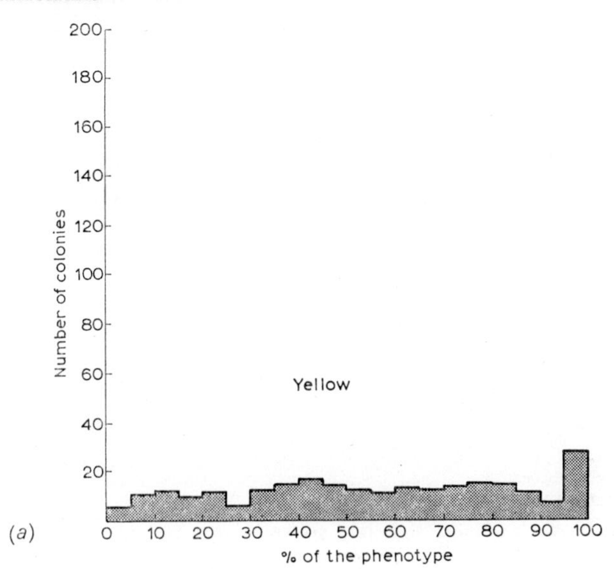
200
180
160
140
120
100
80
60
40
20
Number of colonies
Yellow
(a)
0 10 20 30 40 50 60 70 80 90 100
% of the phenotype

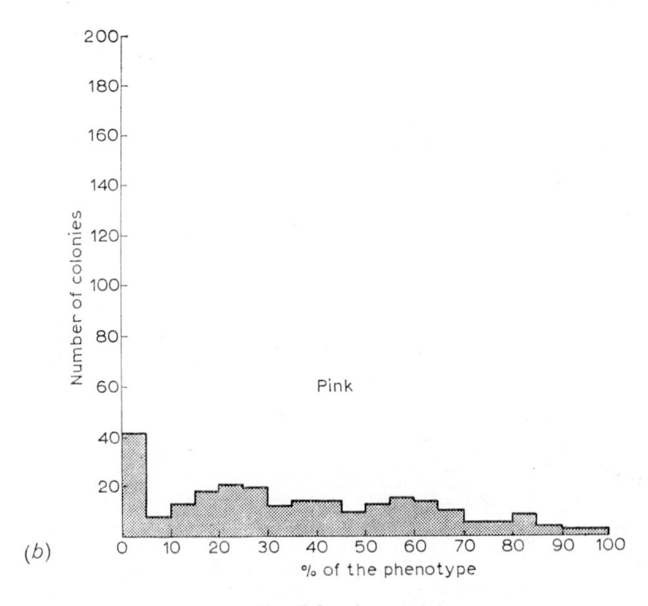
200
180
160
140
120
100
80
60
40
20
Number of colonies
Pink
(b)
0 10 20 30 40 50 60 70 80 90 100
% of the phenotype

Fig. 7.2 (*a*) and (*b*)

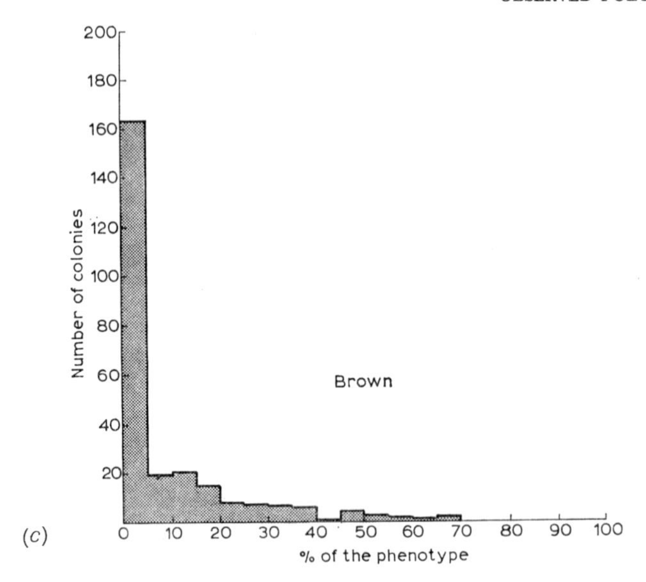

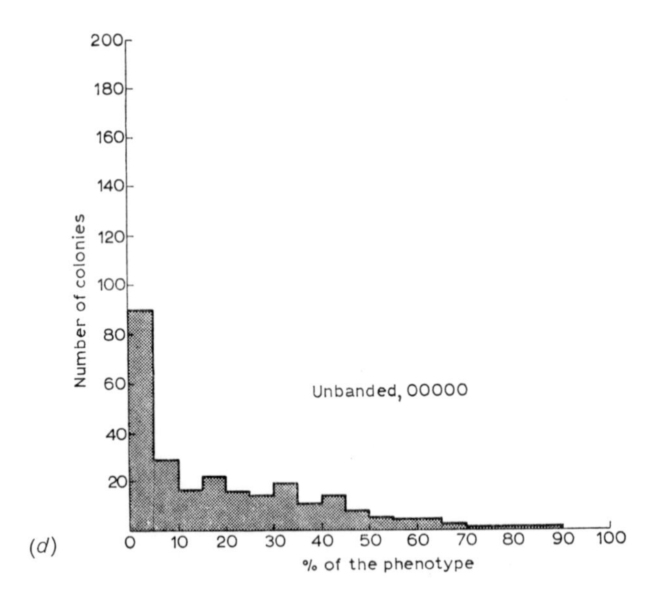

Fig. 7.2 Frequencies of six phenotypes (**a**-**f**) of *Cepaea nemoralis* in 261 colonies on the Berkshire Downs, England. Data from Carter (1968). Yellow, pink and brown refer to the ground colour of the shell 00000, 00300 and 00345 are banding patterns.

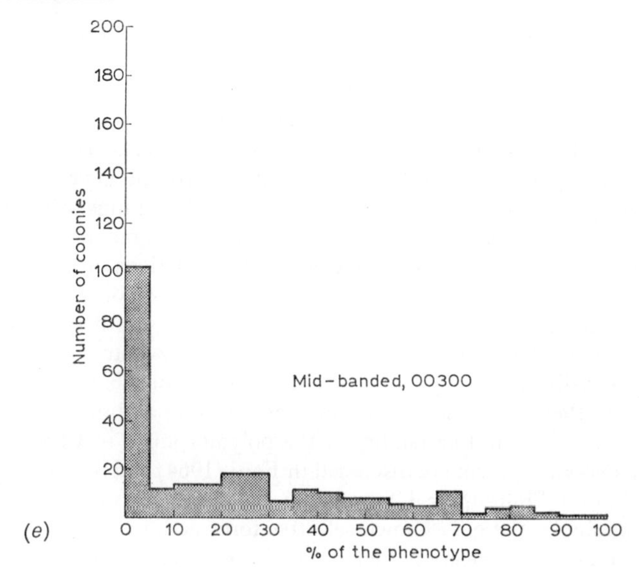

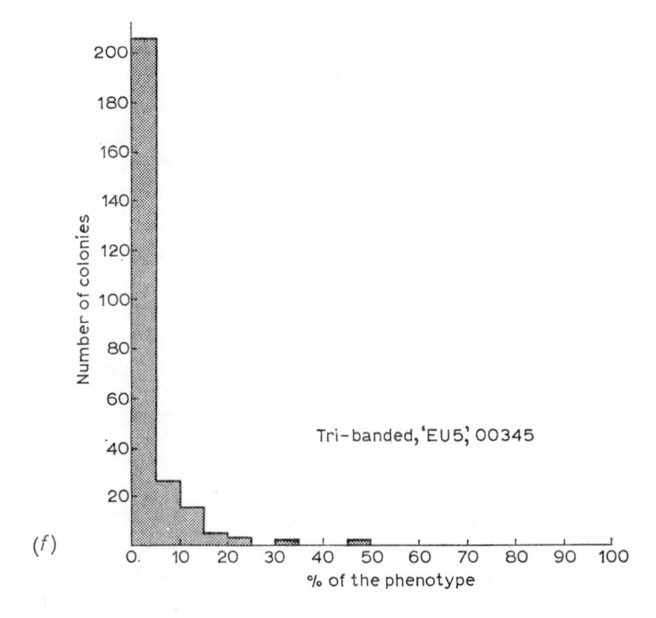

frequencies of different alleles in different populations vary very widely, and this poses a major problem in explaining the maintenance of this polymorphism. The observed frequencies of some of the major forms in some English populations are shown in Fig. 7.2. Heterozygote advantage, the form of selection which is best known to maintain a polymorphism, has often been suggested as the cause in Cepaea. Cain and Currey (1963a) have analysed this possibility, to see what strength of selection is compatible with the wide range of stable frequencies seen. As might be expected from the discussion of *Drosophila pseudoobscura* of the very shallow slope of Δq with heterozygote advantage at extreme gene frequencies, they show that selective forces must be rather small, if all the range of equilibrium frequencies observed for say, yellow, are stable. It may be noted that all these forms in Cepaea show complete dominance in wild populations, in the sense that there is no certain known way of scoring any of the heterozygotes, though small differences between heterozygotes and dominant homozygotes may be found in individual broods in the laboratory. So if there is heterozygote advantage, it is unrelated to the visible phenotype of the shell, which is somewhat surprising. On all these grounds there is a clear case for looking for other and more satisfactory explanations of stability and variability of the polymorphism in Cepaea. Other features of this polymorphism are discussed in Ford (1964) and in a set of papers in volume 253 of the Philosophical Transactions of the Royal Society, 1968.

Rather surprisingly, the only known selective force acting on Cepaea populations, and that includes selection that might be responsible for heterozygote advantage, is that of predation, particularly by thrushes (*Turdus ericetorum*) in the vales of southern England, in the extensively cultivated river valleys, with small wood lots of only a few hectares. Cain and Sheppard (1954) Currey, *et al.* (1964) and Arnold (1970) showed that the effect of thrush predation was to make the general polymorphism of the snails match the background. In all these cases thrushes merely appear as a disturbing factor, changing the equilibrium point of the polymorphism, and there is no suggestion in these studies that they produce the polymorphism. Arnold in preparation has indications, on slightly higher ground in the Chilterns, that the intensity of selection is proportional to the intensity of predation. In many areas thrush predation is slight or absent, and indeed there is some suggestion that many thrushes do not like snails.

In more upland Britain, particularly in the chalk hills, appreciable areas of countryside show consistent frequencies, which then change a few miles further on to different sets of frequencies. This phenomenon has been called 'area effects' by Cain and Currey in 1963, and is shown more fully by Carter (1968). In very low, damp places in Cambridgeshire and in the Netherlands (Goodhart 1962, Wolda 1969a) the area effects become what one might almost call microarea effects, with sharp and unexplained changes in gene frequencies over a few hundred metres. Some of these variations in phenotype frequencies are relatable to physical factors, such as cold air drainage or aspect (Cain, *et al.* 1969, Arnold 1969), but most of them have not been. It is perhaps relevant to say that the soil is much less distrubed on these chalk uplands than in the fen like lowlands of Cambridgeshire and the Netherlands, but no relation between soil factors and phenotype frequency has yet been recorded, though it has not been looked for intensively. So this is a polymorphism that varies sometimes in response to predation, sometimes in response to physical factors, and sometimes in response to quite unknown factors.

Although some earlier authors suggested it as an unlikely possibility, Clarke (1962) was the first to emphasise that predation could, by itself and as a single selective factor, produce polymorphism, and he named it 'apostasy'. The essential requirement for this to work is that the predator must select most strongly against those forms which are common in relation to some standard appropriate to that particular habitat. This is certainly a possible explanation for the English vales, but seems much less probable for the English highlands and much of France, which constitute a larger proportion of Cepaea populations. There is in fact predation by thrushes on some of the chalk uplands (Cain and Currey 1968), but there it seems not to be able to move the polymorphism in face of the stronger selective forces, still unknown, that produce the 'area effect'.

Wolda (1963) has suggested that the problem of Cepaea should be attacked from a joint study of population ecology and population genetics, and this is manifestly a sensible suggestion. So far his results have made the situation somewhat more puzzling. For instance Wolda seems to find bigger differences in clutch size and development rates between different localities, involving a number of different phenotypes, than between different phenotypes from different localities. Some behavioural differences between different phenotypes have also been found but again there are suggestions that these differences are not closely related to the major phenotypic differences of a polymorphism, but are determined either by other loci, or possibly even straight forwardly phenotypically by the environment in which a snail has lived.

Evidence for the importance of ecological effects comes from a comparison of *Cepaea nemoralis* and *Cepaea hortensis*. The joint occurrence of these two species have been studied particularly by Clarke (1969) and Carter (1968). Mixed colonies of these sibling species are not very common, but are commonest on limestone and chalk. The distribution of neither species is readily explicably over any area of more than say 50 miles or so, though in smaller areas they may appear to behave consistently. Where one of the two species is common, and it does not matter which one, the other one tends to be confined to woodlands and not occur in what otherwise is the typical habitat for both of them, grassland and rank herbage. There is also the interesting old observation of *C. nemoralis* and *C. hortensis* on sand dunes by Oldham (1929), which is considered more fully in Chapter 12 and discussed by Cain *et al.* (1969). Whether the polymorphism in one species affects the other is disputed between Carter and Clarke, and the interpretation is undoubtedly difficult because of all the other unknown forces affecting polymorphism, but at most it seems that the effect is not a very large one.

Another phenomenon that may be important in Cepaea as there are a large number of loci involved, is the question of co-adaptation (Goodhart 1963, Cain and Currey 1963b). Like fitness, co-adaptation is a phenomenon which is easily defined in relation to an individual. Different genes must be related to each other during development. Some forms of co-adaptation between individuals, such as sexual differences, are also easily defined. Whether co-adaptation has any meaning in populations, other than possibly being another way of describing the peaks in Wright's Adaptive Topography, is not so clear. Still these peaks will certainly be affected by any interactions in selective effects on different genotypes, as the unit of selection is the individual (Milkman 1967). Clarke has developed an interesting model (1966) to show that it is possible that the change in gene frequency may get displaced spatially from the change in physical factors that produces the selective forces, and Clarke and Murray (1969) have shown a case in *Partula* in Moorea in the Society Islands in the Pacific, where the change in gene frequency appears to be related to the occurence of another species a short distance (100 to 400 m) away. With complications like these, and with the great variety of forms, a complete explanation of the polymorphisms of Cepaea is clearly still a long way off.

Genetic load

These four examples show that polymorphisms can differ greatly from each other, and the full discussion in Ford (1964) exemplifies this much more fully. Selective forces are not necessarily constant, and indeed the presence of selection relating to fertilization in Mus, Panaxia and Drosophila suggests that they may frequently be inconstant. The relation of selection to possible population regulating factors in Drosophila and Cepaea suggests that the ecological factors must be allowed for in discussing genetic variability. And most important of all the unit of selection must always be the individual. With these considerations, the suggestion that genetic variability is so great that selection would produce an impossible genetic load, as suggested for instance in the most important paper by Lewontin and Hubby (1966), seems unlikely. Selection may or may not produce a genetic load depending on how it reacts with other population factors (Turner and Williamson 1968), and the common occurrence of genetic variability of the type demonstrated neatly by Lewontin and Hubby suggests that most forms of selection are associated with population regulating factors, that is to say variable selective values, that tend to produce the least load. That is, the problem of load can be used to study heuristically the types of selection, rather than be cited as evidence against selection. Other aspects of the problem of genetic load need not be discussed further here in view of the recent reviews by Wallace (1970), Manwell and Baker (1970) and Moran (1970). Moran exposes particularly clearly the weaknesses in the usual formulations. In the next Chapter, other interactions between population size and selection are considered.

8 Genetic and demographic interactions

Population ecologists have mostly ignored genetics: population geneticists have almost all worked with gene frequencies and ignored questions of population density. Yet it is now clear that, in the same way that each species has its own demographic characteristics, so has each genotype. Methods in population ecology that do not admit of the possibility of studying genetic variation are as limited as studies in population genetics that eliminate the numbers involved. The study of evolution, and particularly the study of small scale and short term changes below the level of the species, invariably has both ecological and genetical aspects.

The processes used in analysing genetical and ecological situations are already similar, and could be more so. In both cases, the analyst wishes to construct a mathematical model, in which the various terms have a clear cut biological meaning and in which the parameters can be estimated by experiment or observation. In both fields, analysis consists of a cycle of first observing and experimenting on a population, secondly interpreting the observations, frequently by studying the rate of change of some major variable, thirdly building a model to explain the observations and their interpretation, and then going back to the start of the cycle to make more observations so that the model may be fitted and tested. The change of the logarithm of the population size and the change in gene frequency could clearly be combined into the study of the change of the logarithm of the gene number, but this has not been done. Nearly all the methods described so far in this book, whether Leslie matrices, k factor analyses, or the various techniques that have been used to study polymorphism could be adapted for simultaneous ecological and genetical work. For instance, genotypes can be included in the Leslie matrix notation (see page 65) while ecological factors alter the whole tenor of the argument about segregational genetic loads (page 90).

The reaction of genetic and ecological phenomena on each other in systems of competing species and in systems of species feeding on each other are discussed in Chapters 11 and 15. Here, where only single species systems are being discussed, two aspects can be looked at. The first is, to what extent have the generalizations above been used in analysis? The second is, are there any phenomena for which the interaction of genetics in ecology are thought to be essential?

Laboratory populations have quite often shown that different genotypes have different ecological characteristics. The study of Dobzhansky *et al.* (1964) on the intrinsic rates of natural increase associated with different gene arrangements in *Drosophila pseudoobscura* has already been mentioned. There has been quite a lot of work with different genetic strains of Tribolium beetles. For instance Sokal and Sonleitner (1968) working with *T. castaneum* showed that the homozygous mutant 'black' differed from the wild type both in survival and fertility. At low densities, these were about the same in both genotypes. Proportional survival declined, more or less linearly, with increasing population density for both. This decline was only very slight for the wild type, but quite steep for black. In constrast, while fecundity fell off with density in both genotypes, the effect was more marked with the wild type. In this case the reciprocal of fecundity changed linearly with density. This contrast leads to a marked polymorphism, except possibly at the lowest densities. Observationally, there would be an interaction between the equilibria in gene frequency and population density. A model of the system, and one is provided by Sokal and Sonleitner, must include both aspects. Nevertheless, the polymorphism would occur at nearly any density because of the inversion of survival and fecundity between the two homozygous genotypes. Here there is an interaction of genetics and ecology, but this does not affect the form of the population model, only the size of the parameters and the detailed position of the equilibria.

Interactions of this sort have been shown, though nearly always much less clearly, in some field data. For instance, Iwao and Wellington (1970 and earlier) have studied in a number of ways the effect of genetic variables on the population ecology of tent caterpillars. The population size of *Panaxia dominula* has been estimated for a period of 30 years (Fig. 1.14) and the factors affecting its polymorphism have also been studied intensively (p. 78), but the effect that population density may have on the selective forces, or the effect that the gene frequency may have on the population size is still debatable (Lees 1970). In *Cepaea nemoralis*, Wolda's (1969a) detailed study of population numbers and gene frequencies leads to no conclusion about the effect the one has on the other.

Coming then to the second question, are there any situations known in which only the interaction of the two subjects can produce a certain phenomenon? The question can perhaps be made more precise by stating how such situations could be detected. If it is postulated that some observations of population density, or age structure, or any other ecological variable depend on genetic variation, then an appropriate test is to maintain a genetically homogenous population. The easiest and most natural way to do this would be to have one which was homozygous at all loci thought to be involved. In such an experimental population the ecological phenomenon should disappear. A case in point are the cycles in numbers of small mammal populations. Several species of voles of the genus Microtus have been reported to have four year cycles. Krebs (1966), following Chitty, suggests that these cycles are caused by genetic changes. If it were possible to set up genetically homozygous populations, this hypothesis could be tested directly. From what has already been said, gene frequencies can be expected to change during the cycle, and the amplitude and frequency of the cycles might well be effected by the genetic constitution of the population, but neither of these would demonstrate a genetic cause for the cycle. Only if cycles occurred only in genetically heterogeneous populations would it be reasonable to assume that genetic variation was essential for the production of cycles.

Conversely, a polymorphism might depend on density effects for its maintenance. In theory, this is quite simple. Any system that can produce stable competition between two species (and these systems will be discussed in Chapter 11), can also produce a polymorphism of two genes (Williamson 1957). That is, if two genotypes have separate density controlling systems, a polymorphism is readily produced. Testing is even less easy in this case, as a polymorphism would be expected to persist if the population density were held constant. Observations such as those by Beardmore (1970) in experiments on the fitness of Drosophila populations, that genetically heterogenous populations had a higher population density than cultures of homozygotes, suggests that density dependent selection (Turner and Williamson 1968) is involved. The complexity of the polymorphism in Drosophila was discussed in Chapter 7, and it is hardly likely that density dependent selection will be the only selective force involved in the maintenance of the polymorphism.

Ecologists can probably accept the importance of knowing something about at least the major genetic variation in a population quite readily. The problem for population geneticists is rather harder. As is noted in Chapter 6, there is really no indication of how commonly selection varies with frequency or density. If ecological factors are often of importance in the analysis of genetic systems, then density dependent selection would be likely to be quite common, and constant selective forces might be rather rare. Even so, the variation in the selective coefficient might be so small that the theory based on constant selection would still be adequate. This problem is undoubtedly one that needs study, and almost the whole of the theory of population genetics has been based on constant selective values.

III Two species interactions

9 The nature of the problem

Two species can interact in a great number of ways, and the consequences of these for the two populations are also varied. Indeed, remembering that in most species each individual is apparently genetically different from all others, the effects of the interaction may well be appreciably different for different individuals or different genetic types. There is a natural tendency for biologists to want to classify interactions by the biological processes involved: if an individual of one species bites one of another that is one type of interaction; while if it gets in the way of the other's reproduction, by interfering with egg laying sites for instance, that is another. However, if we pause for a moment and consider the great variety of possible interactions: the chemical interactions that probably occur between bacteria and between planktonic algae (Fogg, 1965), the variety of ways that fungi attack other organisms, the fungi and algae in lichens, the questions of shading and soil development involved in plant succession, and particularly the astonishing variety of ways of life among animals, and the remarkable ways of getting a meal, then the hopelessness of this approach can be seen. A simple classification would be naive, easily confounded with examples, while a comprehensive one would be quite unwieldly.

Symbolic definitions

The simplest way out of this difficulty, and the natural one in population analysis, is to classify interactions not by their actions but by their effects. This seems to have been first done systematically by Odum (1953). He classified effects by calling those that the population of one species has on another, positive, neutral or negative depending on whether the second species increases, is unaffected by, or decreases in the presence of the first. Although he was not explicit on this point, from the context he was considering equilibrium population densities, or, what would nearly always come to the same thing, the sign on the appropriate term in the differential equation describing the populations.

Thus if species G has a positive effect on species F, the equation dF/dt must include a term in which G, or a monotonic increasing function of G, has a positive sign

$$\text{e.g. } dF/dt = rF + sG^{\frac{1}{2}} - qH^3,$$

where F, G and H are population densities of three species, r, s and q constants. H has a negative effect on F.

Effects of species F on G	Effects of species G on F: +	0	−
+	++	+0	+−
0	0+	00	0−
−	−+	−0	−−

The central element of this table where neither species interacts with the other, is of no interest. Because of symmetry, the remainder of the table fits into five categories:

++ symbiosis
+0 commensalism, etc.
+− predator/prey, including parasite/host, herbivore/plants etc.
0− amensalism
−− competition

The first of these is a fascinating biological topic, but its importance in populations in general is small, and I shall not discuss it further. The second is probably very important, and includes all those cases where one species produces conditions which are a necessary prerequisite for the existence of another. For instance, the duckweed *Lemna trisulca*, is found around the roots of *Lemna minor*. *Lemna trisulca* requires a carbohydrate which is given off by other plants (Landolt 1957). Again Powell (1958), as was noted in Chapter 4, recorded *Pseudomonas aerugionosa* as a constant contaminant in a culture of *Aerobacter cloacae*, apparently living on the compounds produced by Aerobacter. Nevertheless, this type of interaction has not been studied with a view to analysing populations. It is possible that, considering the effects on the population and not on the individual, many predator/prey interactions are of this type. The predator population clearly gains from the presence of the prey. It is possible that in some cases prey populations are not affected by the predator; because all the casualties to predation would have died soon anyway, or because deaths from predation are compensated by lower mortality from other causes. This will be discussed further in Chapter 13.

The remaining three interactions are those normally studied by ecologists. It is interesting that the predator/prey interaction, which is easy to observe in the field, has not lead to many satisfactory laboratory experiments, while there have been many laboratory experiments on competition although the occurrence of competition in nature is a subject of dispute. The main point requiring thought at this stage is the distinction between amensalism and competition. Using this definition, competition is a two way process. However, it is frequently used, particularly by botanists and also by marine biologists, for systems in which only one species is affected, which here I have called amensalism. For instance Skellam (1951) made an interesting model of two hypothetical plant species. One is superior in dispersal, the other in growth, at a particular site. His equations for the equilibria are

$$G_1 H_1 N_1 - \log(1-N_1) = 0$$
$$G_2 N_2 \log(1-N_1) = G_1 N_1 \log(1-N_2/1-N_1)$$

where N_1 and N_2 are the population densities, G_1, G_2 reproductive parameters for the two species, H_1 a habitat parameter for the first species. It can be seen that the second species, the quick disperser, does not affect the first. This system leads to a stable equilibrium with both persisting. It is an interesting model for Hutchinson's (1953) fugitive species, and possibly even for Salisbury's (1942) species of transient habitat, but it is not a model for competition as defined here, but of amensalism, or, as it might be called in this case, suppression.

Verbal definitions

Many ecologists do not accept this type of definition, usually prefering one which does not use mathematical terms. Milne (1961) for instance, argues strongly both for a definition based on individual effects, and one couched verbally not mathematically. He was concerned only with the definition of competition, but no doubt would apply the same criteria to other interactions. The study of individual actions is a necessary part of the study of any population process, though the detail needed will not always be very fine; but it can only cause confusion to set up such studies as alternatives to studying population effects. It is certainly legitimate to call certain individual interactions 'competition' and then to study their population effects, but it does not follow that such studies will necessarily cover the same phenomena as 'competition' labelled by population effects in the first place. All the interactions discussed in this book are classified by their effects on the population concerned.

I have given two reasons so far for not classifying ecological interactions by their effects on the individual: their great variety, and the possibility that it would cause confusion with similar phenomena in the population. The third reason is related to these: populations must be discussed and analysed in quantitative terms, and so definitions are necessarily to some extent mathematical. Those who think in terms of individuals seem to prefer verbal definitions, which on examination are quite insufficiently precise for quantitative study. For instance Milne (1961) defines competition as 'the endeavour of two animals to gain the same particular thing or to gain the measure each wants from the supply of a thing when that supply is not sufficient for both'. This is to substitute for one undefined term, competition, seven, namely endeavour, gain, particular thing, measure, wants, supply, and not sufficient. That such an unusable definition as this should be widely quoted shows only too clearly how little quantitative work has been done on population interactions. Ecologists who are attracted by such a definition should consider how one could decide in any particular case, such as the beetles *Tribolium confusum* and *T. castaneum* cultured together in the same vial of flour, whether or not they are in competition. This example will be considered more in the next Chapter, but it can be said here that the two species are clearly in competition by Odum's definition, but that in as far as Milne's definition can be interpreted at all in this situation, they would appear not to be competing. Yet this system is probably the most famous experimental competition system there is.

Applying a definition to population analysis

Testing is the critical stage in applying a definition, or to put it another way, a definition is only useful if, in a real biological situation, it is possible to apply a test for a phenomenon so defined. Even with Odum's definition, this is not always straightforward if real species are discussed rather than species F and G. For instance Broadhead and Wapshere (1966) studied two species of Mesopsocus (*M. unipunctatus* and *M. immunis*) on Larch twigs. These are psocopteran insects, which feed on the algae that grow on twigs. These two species are attacked by two hymenopteran parasites, a Myrmarid (*Alaptus fusculus*) attacking the eggs and a Braconid (*Leiophron* cf *similis*) attacking the sixth instar. Their model is set out in their Table 26, which is given here as Table 9.1. There is density dependence in periods 2, 6, 7 and 8. In all other periods a constant fraction, labelled survivorship or sex ratio, is assumed to pass on to the next period.

It is perhaps not immediately obvious whether or not, in terms of this model, the two species of Mesopsocus are in competition. They are linked both by the Myrmarid and Braconid, but are otherwise independent. If the numbers of one species are increased where should we look for a decrease in the other: in all stages or in any one stage? And how do we increase the equilibrium numbers of one? Manipulation of the model shows that, in this case, the questions are fairly simple to answer. Remembering that any death rate tends to decrease the equilibrium population size, this can be changed for either species by adjusting survivals. If the survival at stage 5, after the Myrmarid attack and before the Braconid attack, is doubled first in one species then the other, the population changes in the way shown in Table 9.2.

Table 9.1 A mathematical model for two species of Mesopsocus, and two species of parasitoid attacking them, from Broadhead and Wapshere 1966

The parameters of the model (numbers per 16 twig sample unit)

Period		Factor	Parameter	*M. immunis*	*M. unipunctatus*
1. Eggs laid	→ Eggs available to mymarid	Fungal attack	Survivorship ($n = 5$)	0.879	0.955
2. Eggs available (u_I)	→ Eggs surviving (u)	Mymarid attack	Area of discovery (a) ($n = 4$)	0.003096	0.01656
3. Eggs surviving mymarid attack	→ Nymphs within eggs at beginning of winter	Attack by sucking predator	Survivorship ($n = 4$)	0.879	0.776
4. Nymphs within eggs at onset of winter	→ Nymphs ready to hatch in spring	Winter loss of egg batches from twigs	Survivorship ($n = 3$)	0.612	0.974
5. Nymphs ready to hatch from eggs in spring	→ 5th–6th instar nymphs available to braconid	Predation by birds and lacewings	Survivorship ($n = 4$)	0.185	0.267
6. 6th instars available (u_I)	→ 6th instars surviving (u)	Braconid attack	Area of discovery (a) ($n = 3$)	0.1807	0.0653
6th instars surviving	→ ♀ 6th instar nymphs (N_0)	—	Sex ratio (♀ : total)	0.597	0.506
7. N_0	→ Females at end of preoviposition period (N_I)	Mortality + migration	Regressions below		
8. N_I	→ Eggs laid	Oviposition	Regressions below		
		PARASITES:			
Mymarids produced by oviposition in host eggs	→ Mymarids attacking the following year (p)	Mortality of host eggs: from sucking predator	Survivorship	0.879	0.776
		and by winter loss of host egg batches	Survivorship	0.612	0.974
Braconids produced	→ Braconids attacking the following year (p)	—	Survivorship	1.0	1.0

EQUATIONS USED:

	M. immunis	*M. unipunctatus*
7. Period $N_0 \rightarrow N_I$	$\log_e N_I = 0.809 + 0.364 \log_e N_0$	$\log_e N_I = 1.323 + 0.305 \log_e N_0$
8. Period $N_I \rightarrow$ Eggs	$\text{eggs} = 16.16 + 28.38 \log_e N_I$	$\log_e \text{eggs} = 3.564 + 0.694 \log_e N_I$
2 + 6. Parasite attack	$u = u_I e^{-ap}$	

The results are slightly unexpected at first sight, because increasing the number of Mesopsocus available to Braconids has chiefly benefitted the Braconids. Still, the Mesopsocus species are clearly interacting, and allowing for the change in Braconid numbers, are tending to depress each other. So in this model, the two species are in slight competition. Broadhead and Wapshere incidentally concluded (p. 380) from their field data that, while there may have been slight competition, most of the 'findings point to the lack of any competitive interaction of importance'.

Table 9.2 Numbers of Mesopsocus spp. per 16 twigs predicted by Broadhead and Wapshere's (1966) model

	Eggs	Nymphs	Adult females
Original model			
M. immunis	40.2	16.1	2.3
M. unipunctatus	118.9	36.6	5.8
Braconids			2.6
immunis survival improved			
M. immunis	34.5	14.3	1.9
M. unipunctatus	111.5	39.4	5.2
Braconids			10.9
unipunctatus survival improved			
M. immunis	13.6	5.4	0.9
M. unipunctatus	123.8	36.9	6.1
Braconids			8.8

This conclusion is both reasonable and in agreement with their model. The species are not in strong competition, and the sort of weak competition in the model is normally undetectable and ignored in field studies. They also note that their model suggests that the two parasites are in competition, as changes in the parameters leading to an increase in one produce a decrease in the other and vice-versa.

In the rest of this Section, various ways of applying Odums' definitions to populations will be discussed. Chapters 10, 11 and 12 deal with competition starting with some laboratory experiments, clarifying the theory and discussing the evidence of competition in natural populations, including experiments with natural populations. Chapters 13 and 14 deal with predatory/prey interactions. The Section ends with a discussion, in Chapter 15, of some of the genetical consequences of predatory/prey systems. The genetical consequences of competition are discussed in Chapters 11 and 12.

10 Competition experiments

A variety of organisms has been used in the laboratory for what are called competition experiments. Usually two closely related species are chosen and grown both separately and together under carefully controlled conditions. The population or part of it, is counted at regular intervals. At the end, the changes in the single and mixed species replications are described, and the experiment analysed. The analysis is usually partly in verbal terms—whether the species are in the same niche—and partly mathematical. The concept of a niche is discussed in the next Chapter. The mathematical part sometimes involves comparing the course of the experiment with some theoretical formulation.

Animals that have been used in such experiments include protozoa, particularly Paramecium; a number of beetles, particularly Tribolium spp.; the fly Drosophila; and fresh water cladocera, particularly Daphnia spp. Experiments with microbes other than protozoa have been rare, though a number are reported by Crombie (1947). Experiments like these have hardly ever been done with plants, though Clatworthy and Harper (1962) have used Lemna, the duckweed, a small floating angiosperm. After describing and discussing some of the classical experiments, other experiments and techniques, more related to the study of agriculture, are considered at the end.

Gause's experiments with Paramecium

One of the most famous series is Gause's (1934) with *Paramecium aurelia* and *P. caudatum*. These are described in Chapter 5 of his book, and his counts given in his Appendix 3. The experiments have been reanalysed several times, that by Leslie (1957) being by far the most useful. Most textbooks of animal ecology also describe these experiments, but usually incompletely and frequently incorrectly. There are two difficulties in working with Paramecium that are not mentioned by Gause or Leslie. The animal has two nuclei, a diploid micronucleus and a polyploid meganucleus. The meganucleus has to be formed again from micronuclei from time to time either by conjugation, in which micronuclei are exchanged between a pair of animals, or by autogamy, which takes place within one animal. Both these processes delay division considerably. If they are prevented, the animals show senescence and die (Sonneborn 1954). The other difficulty is not a difficulty for the experimenter, but only for his successors. Geneticists have shown the classical species of *Paramecium* (*aurelia, caudatum, bursaria* and five others) consist of a number of 'syngens' which, as Hairston (1958) points out, are perfectly good species and merely lack conventional names.

There are at least fourteen of these for *aurelia.* As these were not known when Gause did his work, we do not know what species (syngens if you will) he did in fact use.

The experimental system used was described in Chapter 3, but may be briefly restated. The Paramecia were grown in 5 cm³ of Osterhaut's medium, a simple mixture of salts, at 26°C in test tubes. Each day they were resuspended in fresh medium and given fresh food. The food was the pathogenic bacterium *Pseudomonas aeruginosa* grown on agar plates. A platinum loop was used to give, as nearly as possible, the same amount of food each day. *P. aeruginosa* does not grow in Osterhaut's medium. The experiments were started with 20 Paramecium of each species on day 0, not counted on day 1, and thereafter counted every day until the experiment was ended on day 21. One such experiment, the one analysed by Leslie, is shown in Fig. 10.1. If grown separately, both species have S-shaped

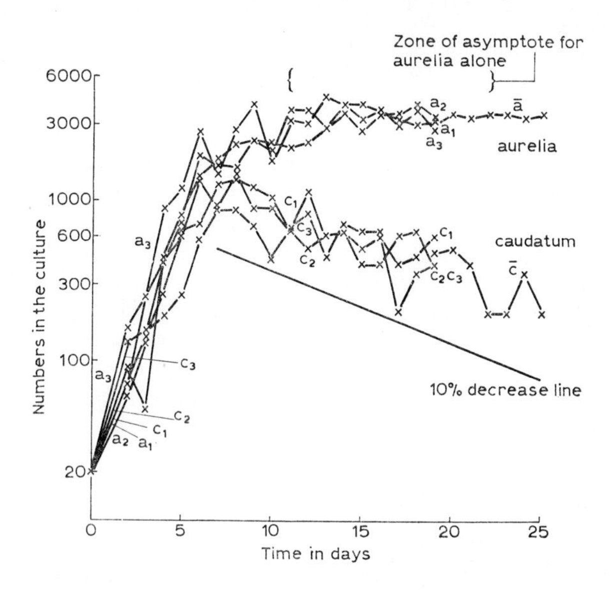

Fig. 10.1 Competition between *Paramecium aurelia* and *P. caudatum* in Osterhaut's medium. Data from Gause (1934) Appendix Table 3. The numbers for three replicates, marked a_1, c_1 etc, are given up to day 19, and for the mean of the three thereafter. Note the zone of the asymptote for *P. aurelia* alone, from Fig. 3.5, and that there are still more than 10 times more *P. caudatum* at the end of the experiment than there were at the beginning.

growth curves (plotted arithmetically) and come to a fairly well defined asymptote (see Fig. 3.5). Gause, Leslie and most other commentators have fitted logistic curves to these data, and some of the limitations of doing this have been discussed in Chapter 3. Grown together though, *P. caudatum* starts well, but its population declines from day 8. *P. aurelia* reaches what appears to be an asymptote about day 10, though this is at a population density appreciably less than that in the single species culture. There is no doubt that each species has affected the other, and that there has been competition.

As there is a constant, limited quantity of food, this experiment is frequently described as simply one of food competition. This is correct as far as it goes, but misses a major feature of the experiment. Leslie fitted an elaboration of the logistic equation to the system:

$$d\ ln\ n_1/dt = a_1 - b_1 n_1 - c_1 n_2$$
$$d\ ln\ n_2/dt = a_2 - b_2 n_2 - c_2 n_1$$

where n_1 and n_2 are the numbers of the two species and a_1, a_2, b_1 b_2, c_1, c_2 are constants. He used Volterra's simplification of $c_1 = b_2$ and $c_2 = b_1$, so that the species differ only in the intrinsic rates of increase, the parameters a_1 and a_2. These equations imply, because of the terms with c, that an increase of one species will lead to a decrease of the other. This indeed seems to have occurred in Fig. 10.1, the number of *P. caudatum* do decline.

There are two unsatisfactory points about this formulation. If the *aurelia* are so much more efficient at capturing bacteria that the *caudatum* get none at all, which is rather unlikely, there is still no reason why the *caudatum* should decline in number. The *aurelia* do not attack *caudatum* or produce substances that kill *caudatum*, and *caudatum* does not die from starvation as quickly as that. If *caudatum* were simply being out-competed for food, its number should neither increase nor decrease. Indeed Gause showed a 'stalemate' between two competitors using the yeast *Saccharomyces cerevisiae* and *Schizosaccharomyces kephir* (=*S.pombe*) in Chapter 4 of his book. In mixed cultures both species rose to an asymptote, but more slowly and to a lower value than when grown singly. By itself, Saccharomyces went to 240 cells per square of the counting chamber, but only to 170 in competition, while for Schizosaccaromyces the figures were 270 and 60.

Why then did the *caudatum* decline? The answer has already been mentioned in Chapter 3, and is given by Gause. As he says 'In our experiments the principal factor regulating the rapidity of this movement of the population that had ceased growing was the following technical measure: a sample equal to $\frac{1}{10}$ of the population was taken every day and then destroyed. In this way a regular decrease in the density of the population was produced and followed by the subsequent growth up to the saturating level to fill in the loss.' In fact *aurelia* was successful in competition, because near the equilibrium point it was still increasing by 10% per day, while *caudatum* was only increasing by about $1\frac{1}{2}$% as can be seen to some extent on Fig. 10.1, by comparing the *caudatum* lines with the 10% decrease line. It can be seen too that *caudatum* had not been eliminated at the end of the experiment: there were still appreciably more than there had been at the beginning. The failure of *aurelia* to reach its expected numbers can reasonably be ascribed to the presence of *caudatum*.

The critical factor in this type of competition is the ability of a population to seize enough food at high densities to compensate for the experimental removal of part of the population. With the logistic model used by Leslie, this is the same as having the higher intrinsic rate of natural increase, but biologically this need not be so. If the birth rate of a species A declines more rapidly with increasing density than that of species B, A may have the higher intrinsic rate, at negligible densities, but the lower rate at the densities involved in competition. This is the second of the two reasons noted above for thinking that the conventional analyses of Gause's experiment are unsatisfactory.

The logistic equations used to describe the course of events in Gause's experiments are not analytic. Without a clear distinction of terms for birth and death in which the 10% death rate would be obvious, they are bound to be misleading. Thirty-five years after the experiments were done, and with our greater knowledge of the biology of *Paramecium*, it is not worth formulating better equations, as the main features of what happened are clear. Gause's (1935) later experiments, in which *Paramecium bursaria* and *P. caudatum* competed without either eliminating the other, have never been fully analysed.

More complex competition experiments

Paramecium is a fairly simple organism. All the individuals of a culture are more or less the same, and when autogamy and conjugation are prevented, then reproduction is simply by binary fission. Most animals have more complex life histories. For instance beetles that live in flour or grain have a life history of eggs, larvae, pupae and adults. Park's famous experiments on competition have used two such species, *Tribolium confusum* and *T. castaneum*. His normal technique was to grow the beetles in 8 grams of flour, and every 30 days to sieve out all stages of the beetles and resuspend them in fresh flour. A typical life span is: eggs 4 days, larvae 18 days, pupae 6 days, immature adults 6 days, and adults several hundred days. Food is not limiting, because it is always renewed, and the principal interaction is apparently cannibalism. The larvae and adults attack chiefly the eggs and pupae, and also to some extent other larvae and young adults. This cannibalistic habit is discussed in detail in Park, Mertz, Grodzinski and Prus (1964).

In Park's experiments with these two species of Tribolium, one always eliminated the other though the time taken for this varied from 180 to 1480 days. Lerner first showed that the outcome depended on the particular genetic strains used, and Park, Leslie and Mertz (1964) described some strains of *T. castaneum* that always lose to *T. confusum* and others that always win. Crombie (1947) managed to maintain *Tribolium confusum* and another beetle *Oryzaephilus surinamensis* in stable competition, but not in plain flour, where *O. surinamensis* was eliminated. In grain, or in flour to which pieces of fine glass tubing were added, *O. surinamensis* persisted. It seems that the extra protection from predation by Tribolium was critical. There is a possible comparison here with Leslie and

Gower's (1960) simulation of protection in a predator-prey system, discussed in Chapter 14.

The full analysis of the experiments with beetles would require a complex population model, preferably based on an extension of the Leslie matrix form. The closest approach to this seems to be Landahl's (1955) analysis of the experiments made by Strawbridge in Park's laboratory. Strawbridge counted his beetles every three days, and thereby not only got more detailed information of the change in population size, but an appreciably higher population density. Landahl's model is based largely on cannibalism and gives a reasonable immitation of the main features of Strawbridge's data. As the immature stages of the two Tribolium species cannot be distinguished, it is difficult to determine all the cannibalistic rates, but it seems that an adequate analytical model is possible in principle, even in these complex cases.

These laboratory experiments show that the course of competition may depend on both the detailed physics of the system and on unexpected biological features, that the elimination of one species by another is a slow process, and that states of persistent competition are apparently possible. Experiments designed to be analytic have been rather scarce.

Replacement series

Competition between crops has long interested agriculturalists. This has lead to some elegant and analytic experiments with plants, using the replacement series techniques devised by de Wit (1960), and elaborated in later papers such as de Wit, Tow and Ennik (1966). The technique could well be applied to microbes and animals as well, and has been used with Drosophila by Seaton and Antonovics (1967). In its simple form, it consists of sowing, at the same density, both pure populations and various mixtures of two possible competitors. To that extent it is comparable to the usual methods in laboratory competition experiments with animals, but it is more closely controlled, in that all the sowings are at the same density, and in that the analysis is simultaneously of the single and mixed species experiments.

Plant populations frequently suffer from the disadvantage, for both experiments and observations, that individuals live a long time, as is shown in Figs. 1.12 and 1.13. de Wit's method circumvents this by considering only what happens to the first generation. However, the results can be interpreted to show what would happen over a series of generations, provided the density were kept the same.

The production of the two species can be measured in various ways, such as the total dry weight produced, the nitrogen yield and so on. If the measure used is the number of seeds produced, then, if the parent plants are removed each generation, the replacement series could indicate the course of a competition experiment. Figure 10.2, from Harper (1968) shows results with two varieties of *Linum usitatissimum* at different densities. If the lines on the figures are convex

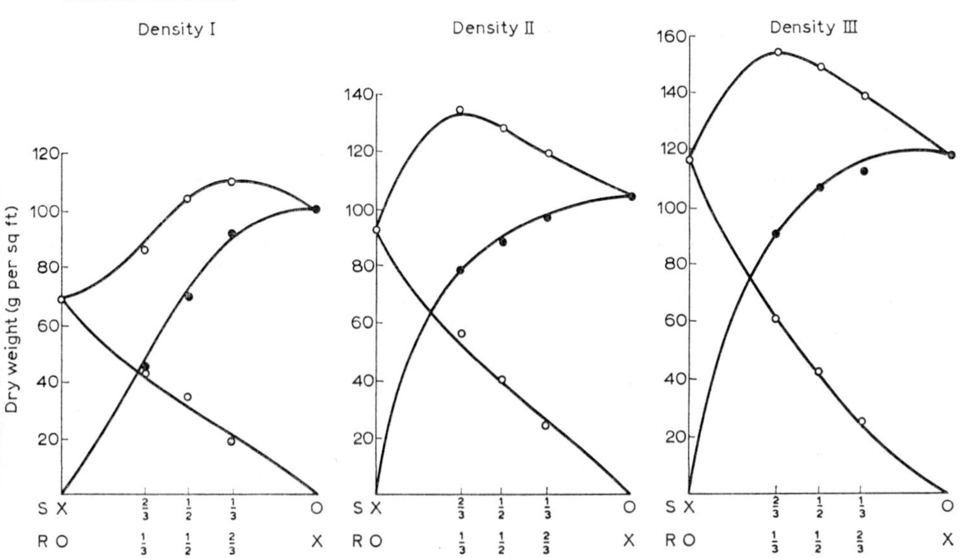

Fig. 10.2 Replacement series diagrams for oil seed (R) and fiber flax (S), two varieties of *Linum usitatissimum.* The x axis shows the proportions of the two varieties sown, the y axis the yield in dry weight, in grams per square foot at 119 days. The upper line of the graph gives the combined yield, the lower, crossing, lines the yield of the individual varieties. It can be seen that, at all densities, the yield of the mixture is greater than the yield of fiber flax, and usually than the yield of oil seed. In all proportions and at all densities oil seed produces proportionately more and fiber flax proportionately less than would be expected from the seed sown. (from Harper 1968).

with respect to the x axis, the species is performing better when it is a minority in the mixture than would be expected from the pure stands. Conversely, lines curved down towards the x axis, indicate poor competitive ability. Seaton and Antonovics, using the replacement series with Drosophila, were able to show a change in the shape of the diagram towards convexity in only four generations. Although their experiment has some unnatural features, they make the important point that changes are more towards the avoidance of competition than towards the improvement of competitive ability. In other competition experiments, this sort of selection could happen accidentally, and, as is seen in beetles, the genetics of the strain used may always be important. The difficulty in allowing for this is that the genes involved are usually minor genes, polygenes, and so not easily scored.

Diallel analysis

All the experiments described so far have been between two competitors, and the difficulties of analysing even these are considerable. Nevertheless an extension to several species is possible using Norrington-Davies' (1967) adaptation of the genetical technique of diallel analysis. This is simply to grow the competitors in all possible pairs and then to use the analysis of variance. As with the replacement series, this is primarily intended to predict results in agriculture, but it might be more generally useful. At least it would encourage a more statistical study of competition experiments.

Competition experiments are more complex than might be expected from their simple design. The results show the need for a sound theory of competition, for the elaboration of more complex mathematical models, and for accurate and detailed descriptions of both the physical and biological features of each experiment.

11 The theory of competition: ecological and genetical effects

The previous two Chapters have dealt with the problems of defining competition and studying it experimentally. In this Chapter, the possible theoretical outcomes of competition will be considered, including both the straightforward ecological results and the more subtle genetic changes that might be expected. Some comments will also be made on the concept of the ecological niche.

The possible outcomes of competition are most easily considered with diagrams showing the combined populations of the two species as a single point in two dimensions. One axis gives the number of one species; the axis at right angles, the number of the other. Given the equations for the change in number of both species, vectors can be drawn showing the expected change from any given joint population.

As in continuous cultures, it is convenient to consider the equilibrium of each variable separately. The technique is most easily demonstrated using the elaboration of the logistic equation:

$$d \ln n_1 / dt = a_1 - b_1 n_1 - c_1 n_2$$
$$d \ln n_2 / dt = a_2 - b_2 n_2 - c_2 n_1$$

Even this formulation has six constants, and as was shown in the analysis of Gause's experiments, can be unacceptable on biological grounds. The first species will be in equilibrium when $d \ln n_1 / dt = 0$, that is when $n_1 = (a_1 - c_1 n_2) / b_1$. The locus of this equilibrium is a straight line with negative slope, as can be seen by the way n_2 comes into the expression. Similarly the locus of the equilibrium of n_2 is another line of negative slope. With n_1 plotted on the x axis, and n_2 on the y axis, four distinct situations are possible, and these are shown in Fig. 11.1. At the equilibrium locus of n_1, the first species is neither increasing nor decreasing, and so the vector showing the movement of the joint population must be vertical, either upwards or downwards. Similarly the vectors on the equilibrium locus of n_2 must be horizontal. If the loci do not intersect, one species will eliminate the other, irrespective of the constitution of the original mixture. The equilibrium densities of the single species are a_1 / b_1 and a_2 / b_2, and it can readily be seen that the species with the locus further from the origin will win. This species does not necessarily have the larger value of a / b, nor necessarily the larger value of a, the intrinsic rate of natural increase.

In Chapter 10, it was noted that Volterra and Leslie made the simplification of $c_1 = b_2$ and $c_2 = b_1$. That makes the two loci parallel. Without this restriction the two lines can cross, and in two ways. In one case there is an unstable equilibrium of the two species, and competition will lead to one species winning, but which

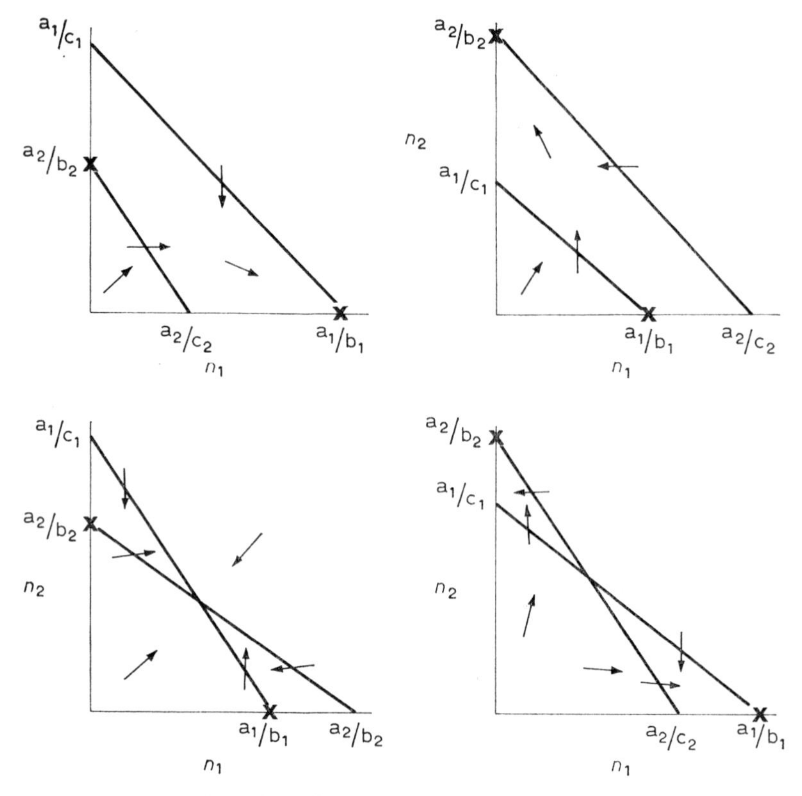

Fig. 11.1 Diagrams of the four simplest competitive situations. The numbers of one species, n_1, are plotted against those of the other n_2. The locus of $d \ln n_1/dt = 0$ lies between the points a_1/b_1, which is the equilibrium of n_1 by itself, and a_1/c_1, and similarly for n_2. The equilibrium for each single species is marked by a cross. The top two figures show competition leading to one species eliminating the other from any initial pair of populations. The bottom left hand figure shows stable competition with both persisting. The bottom right hand figure shows an unstable point, so that only one species survives, but which is eliminated depends on the initial populations.

one will depend on the initial proportions. In the other case, there is a point of stable equilibrium. Both species are depressed from the numbers they would have singly, as might be expected in a negative interaction, but, in this model the joint equilibrium population is greater than either single equilibrium population. With more complex models this need not be so, as a few minutes drawing curved lines on such a diagram will show.

A conclusion that will almost invariably hold in a state of stable competition, is that each species suppresses its own numbers more than those of its competitors, i.e. each is more sensitive to its own intraspecific density. As Pontin (1969) emphasizes, interspecific competition in any normal biological situation implies intraspecific competition too.

These diagrams of competition show the importance of population density in the process. Although it is possible to imagine biological situations in which all the actions of the first species on the second are of a totally different kind from the actions of the second on the first, these are unlikely. If the two sets of actions are the same, for instance using the same food, or as in *Tribolium*, eating each other, then the processes of intra and interspecific competition will be the same, and competition will be in respect of the factors that control the population because of their variation with density (Williamson 1957). This suggests a possible way of detecting the factors that control population size: they will be those involved in interspecific competition.

Ratio diagrams

Another method of showing the course of competition is to plot the ratio of the two species at one time against the ratio at the next time, $n_1/n_2 \mid t_1 : n_1/n_2 \mid t_2$ (Harper 1967, de Wit 1960). On the diagrams used in Fig. 11.1, the change in the joint population size is shown as a vector, at rate of change with a direction, involving two parameters. In a ratio diagram this is shown as a point, needing only one parameter. How this comes about can again be seen in Fig. 11.1. Two coordinates are used, the numbers of the two species, and these are the ordinary rectangular coordinates. However, any other pair of coordinates would do, and an obvious pair is shown in Fig. 11.2. Here the lines radiating from the origin are lines of equal ratio of the two species, and the lines running parallel with

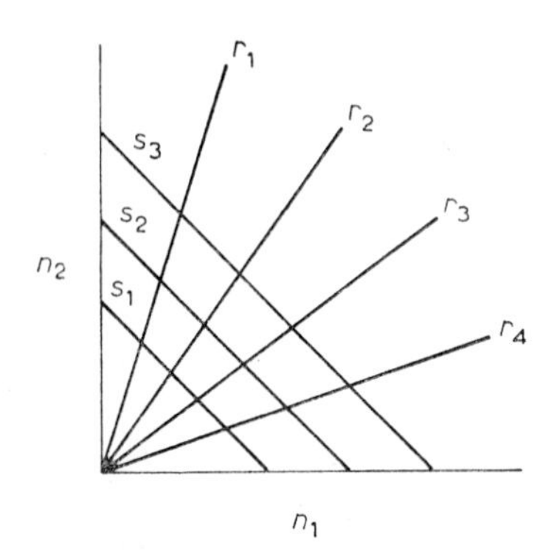

Fig. 11.2 The derivation of ratio diagrams from two species diagrams such as Fig. 11.1, by a change of coordinates. The lines s are loci of constant sums of the two populations, where $n_1 + n_2 =$ a constant. The lines r are loci of constant ratios of the two populations, where $n_1/n_2 =$ a constant. A ratio diagram uses r at one time as one axis, r at another for the other axis.

negative slope are lines of equal joint density. The ratio diagram considers only the first of these coordinates, and, like some forms of the replacement series diagram, ignores the population density. In some situations, as Harper shows, ratio diagrams show the course of competition very clearly, their very simplicity being helpful.

The ecological niche

So far in the discussions in this Chapter and in the two previous ones, it has not been necessary to involve the concept of an ecological niche. There were in fact originally two concepts: Grinnell's of the range of physical environments that a species was found in, and Elton's of a species' position in a community. These were combined to some extent in Hutchinson's and MacArthur's concept of the multidimensional fundamental niche, which has been reviewed by Levins (1968) and by MacArthur (1968). In this, the position of a population is defined in relation to as many axes as the investigator can measure. There is no limit to these and so, as MacArthur says, the fundamental niche has an infinity of dimensions. Elton's niche allows the comparison of species, say, in Europe and Australia, but this feature is lost with the multidimensional niche simply because the dimension Europe—Australia will inevitably separate all the species. Nevertheless MacArthur says that a 'niche', in his sense can be used in a comparative way, in the same way that 'phenotype' is. The analogy seems a little weak. A phenotype can be specified as closely as one likes, usually in terms such as brown eyes, curly wings, while a population can only be specified statistically, e.g. 95% will die in one hour at 40°C.

There is no doubt that MacArthur and Levins have found the concept of a niche useful in developing their theories. There are several difficulties in using the concept in practical analyses of real situations, but only two important ones need be mentioned. The first is that as the number of dimensions is infinite, a selection of dimensions must be used in practice, and there is no clear way of deciding which dimensions to choose. In the past the usage has sometimes suggested a circular definition in relation to competition: if one species eliminates another their niches overlap; if they coexist indefinitely their niches are distinct. Worse, in stable competition the number of each species is depressed, and it could be argued that this necessarily involved overlapping niches. The second objection is that discussions of niches usually imply, and sometimes state clearly, that the limit of a species' niche in each dimension has a clear cut boundary, and even that the distribution is rectangular up to that boundary. This is contrary to experience. The distribution of a species along a dimension is roughly normal, and certainly not rectangular, and there is no definite limit (Whittaker 1970). This can be partly ascribed to the genetical variation amongst the individuals of the population.

Discussions of community structure have perhaps been helped by the concept of the niche, but in discussions of competition it has frequently lead to an avoidance of mathematical thinking, and I think it is good practice to avoid the term niche whenever possible.

Evolutionary considerations

The concept of a niche has also been much used in discussions of speciation, and in the related discussions of the genetic consequences of competition. The importance of defining the genetic strain used was noted when discussing Tribolium. Seaton and Antonivics' (1967) work on Drosophila is a demonstration of how quickly a population can change to avoid competition, and this is the major phenomenon that can be expected. Darwin (1859) discussed the 'divergence of character' and argued that it would be a common consequence of natural selection. Those individuals of a population that avoid competition will usually be at a selective advantage, through a higher birth rate or a lower death rate, and so if the variation is determined genetically, it can be selected for.

The only quantitative theoretical study of this seems to be Bossert's unpublished work, which is outlined and discussed by MacArthur and Wilson (1967) and Wilson (1965). Under certain assumptions, the measurements of two competing species will separate until there is a difference of about three to five standard deviations between their means. That is, competition between the populations will almost, but not quite entirely, disappear. Examples of character divergence of this sort are quite common and are discussed by Mayr (1963). He notes that there are certain cases where character divergence might have been expected, but seems not to have occurred. From the arguments above, divergence can only be expected in characters closely associated with competition, and so with the population controlling factors. Schoener (1965) has made a detailed study of the differences of bill size in different birds, and concludes that the differences are large when the food is sparsely distributed, and in other situations in which there seems likely to be competition for food.

So although a state of stable competition is possible, consideration of the genetical consequences lead to the expectation that, when the biology of the species allows it, there will have been evolution to avoid competition. Putwain and Harper's (1970) experiments, which are discussed at the end of the next Chapter, indicate competition for growing space, and there is clearly no way that a terrestrial plant can avoid needing space to grow in. So for the moment only observations and experiments can determine how often competition occurs in nature, and this will be considered in the next Chapter.

12 Competition in nature

Gause's axiom, or the competitive exclusion principle

Much work in the past twenty-five years has shown that pairs of species that appear at first sight to be in strong competition are in fact not. Some of this work is discussed below. Much of the work on birds started from Lack's (1944, 1947) conviction that Gause's axiom 'two species with similar ecology cannot live in the same region' could be shown to be true in all the cases he had studied. This has been developed into the 'competitive exclusion principle' by Hardin (1960) which says that species in nature are not in competition, and has been criticized by Cole (1960).

This disputed principle clearly depends on what is meant by competition. Here two points are important. The first is that demonstrations like Lack's (which are discussed more fully below) only show that species are not in serious competition rather than they do not compete at all (Mayr 1951). The second is that it is an evolutionary principle. The theory of ecological competition shows that species may coexist permanently in stable competition, as was discussed in the previous Chapter. Laboratory experiments, particularly those of Crombie with beetles, but also Gause's with *Paramecium bursaria* competing either with *P. aurelia* and *P. caudatum* (Gause 1935, 1970) have demonstrated stable competition which have persisted until the end of the experiment. A competitive exclusion principle implies that species have evolved so as to avoid competition. Birch and Ehrlich (1967) seem to think that Lack's studies imply that species have been in strong competition at some time in the past. This is not so. If we see a number of boats sailing along, none touching, we do not assume that this arises from them having touched in the past, but from them never having touched. In this analogy the boats are the species and the water Sewall Wright's adaptive space. Similarly, species managing to avoid strong competition now, will most probably have done so in the past.

The question that the analyst of populations must answer is: is it true, that in nature, if two species are in competition, one invariably ousts the other? and, conversely, that if two similar species occur together permanently, they do not compete? The criterion for competition will be that of affecting each other's numbers downwards. As was shown in the last Chapter, in practice this is the same as sharing controlling factors. Much of Lack's discussion of this topic, and some of the criticism of his work such as Gilbert *et al.* (1952) becomes clearer and simpler if this definition of competition is used. It is perhaps not so surprising that it is a definition that commends itself to Mayr (1963), who is interested in genetics, selection and evolution, but not at all to Milne (1961) or De Bach (1966) who appear not to be. Mayr incidentally has a thorough discussion of competition, with many more examples and references than can be given here.

Once the problem is stated this way, it follows that the only really satisfactory observational tests for the existence and strength of competition would be based on statistics of population size. There are certainly two methods that could be used, and very likely several more. In the same way that *k* factor analysis is used to test for density dependence, as was discussed in Chapter 5, it can be used for testing competition. The other method is based on Principal Component Analysis discussed in Chapter 16. If two species are in competition, there will be a negative correlation between the logarithms of the population sizes. As the species will also inevitably have different habitat preferences and different reactions to variations in the environment, a partial correlation is needed, eliminating important aspects of the environmental variation. This can be done using Principal Components. However, neither method has been discussed in print before, let alone used, and so the discussion of the details and difficulties of these techniques will have to be published elsewhere. In any case, there are not sufficient data to use these methods in discussing the possibility of competition in the field with the other examples discussed in this Chapter. So indirect methods have to be used.

Many have tried to argue for and against competitive exclusion by discussing niches: Gause (1937, 1970) for instance or Crombie (1947). Such arguments are frequently circular. If one species eliminates another they share a niche, if they coexist they are in separate niches. Even this is less unhelpful than Hutchinson (1965), who, by taking suitable axes defines 'The abstract niche is in fact constructed in order to be that thing in which two sympatric species do not live.' The disadvantages of the concept of multidimensional hyperspace niches were discussed in Chapter 11, but this quotation shows it can have no part in answering the question of whether the population of a species is or is not affected by the presence of the population of another species.

Categories of evidence

There are a number of more or less valid categories of evidence showing, indirectly, that species do or do not compete. In the rest of the Chapter, these categories will be discussed in turn, starting with those that most strongly indicate that competition results in the elimination of one of the competitors, and ending with evidence that stable competition is sometimes found. From the discussion in the last Chapter of the genetical consequences to be expected from competition, it is not surprising that exclusion seems to be the normal result of competition; but it is also not surprising that instances can be found in which species seem to be in stable competition.

Displacements

The most direct evidence that one species can displace another is simply that this has been seen to happen. One species invades the country occupied by another and replaces it. There are lots of examples of invasions, many resulting from man's introductions. A famous case, that of the grey squirrel *Sciurus carolinensis* in England, shows the sort of evidence that exists, and the difficulty in interpreting it.

The native English red squirrel *Sciurus vulgaris leucourus* has disappeared over much of England and the grey squirrel is found instead. The spread of the grey squirrel has been mapped and typically the red squirrel disappears ten to fifteen years after the first greys are seen in any particular place. Figure 12.1 shows the

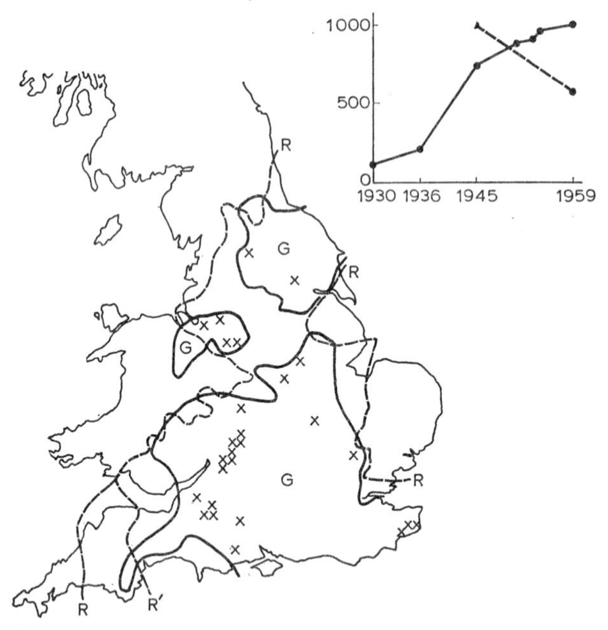

Fig. 12.1 The distribution of red and grey squirrels in England and Wales. The map shows: inside the continuous line, the distribution of grey squirrels in 1945: peripheral to the dashed line, the distribution of red squirrels in 1959 (in the south west continuously to the line marked R, discontinuously to R′): by crosses, isolated occurrences of red squirrels in 10 km squares in 1959, inside the grey's distribution of 1945. The graph shows the records from 10 km squares for each species at various dates. There are 1638 10 km squares in England and Wales. (Map after, and graph from Lloyd 1962).

position in 1959 and the spread of the grey from the two main introductions in Bedfordshire and Yorkshire. Replacement in fifteen years seems about the time one might expect by analogy with laboratory competition experiments for an animal that first breeds as a yearling, and has two litters a year after that, giving only about two daughters per mother per year. (Shorten 1954, Southern 1964).

There are two objections to interpreting this as competition. The first comes from those like Shorten (1954) who want direct evidence of competition and who perhaps expect competition to mean direct interaction between individuals. This is expecting a bit much both of the evidence and the possible nature of the interaction. The second is that there is no evidence that it is reciprocal. That is, the red squirrel may have no effect on the grey. The only possible evidence would be that the spread of the grey is slower than expected. But in this, and in all similar cases, no quantitative study has been made either of the observed rate of spread nor has an expected rate been calculated. A quantitative study of the rate of spread of an immigrant is given by Skellam (1951) but in this case the species concerned, the muskrat *Ondatra zibethica* in central Europe was not displacing anything.

Replacements

The second and third categories of evidence relating to competition in nature can be considered together. They are the evidence from geographical and ecological replacements. As all geographically distinct areas are also distinct ecologically, and as all ecologically distinct areas are also separated microgeographically (Mayr 1947), the categories grade into each other. A well known example of geographical replacement is found in Darwin's finches: *Geospiza conirostris*, which does not occur in the same island as either *G. scandens* (the cactus finch) or *G. fortis* (the medium ground finch). *G. conirostris* comes between the two in its food preferences and is more closely related to *G. scandens*. The distributions are shown in Fig. 12.2. This case can also be regarded as an instance of ecological replacement. The populations labelled *G. scandens* can possibly be regarded as distinct subspecies of *conirostris*, which have evolved so that they are ecologically distinct from *G. fortis*. In Darwin's finches, evolution has primarily been in food habits, and it is reasonable to suppose that food has been the controlling factor for most of the populations, at least while speciation was taking place.

Another mixed example of ecological and geographical replacement is shown in this group. *G. difficilis* is thought by Lack (1947) to be the species closest to the ancestor of all Darwin's finches. Its distribution in the archipelago is almost entirely peripheral to *G. fortis*, but on the central Indefatigable Island *G. fortis* occurs in the coastal scrub land, *G. difficilis* in the central forest. In all these cases, it is plausible, but of course far from certain, that the ecological and geographical distributions remain distinct because of the failure of any genetically divergent members of each species to maintain themselves in the face of competition from the next species. In other words, there appears to be stabilising natural selection caused by competition. Other examples, some involving mixtures of these first three categories, will be found in MacArthur and Wilson (1967).

Character divergence

The fourth category of evidence, that of character divergence, emerges from the

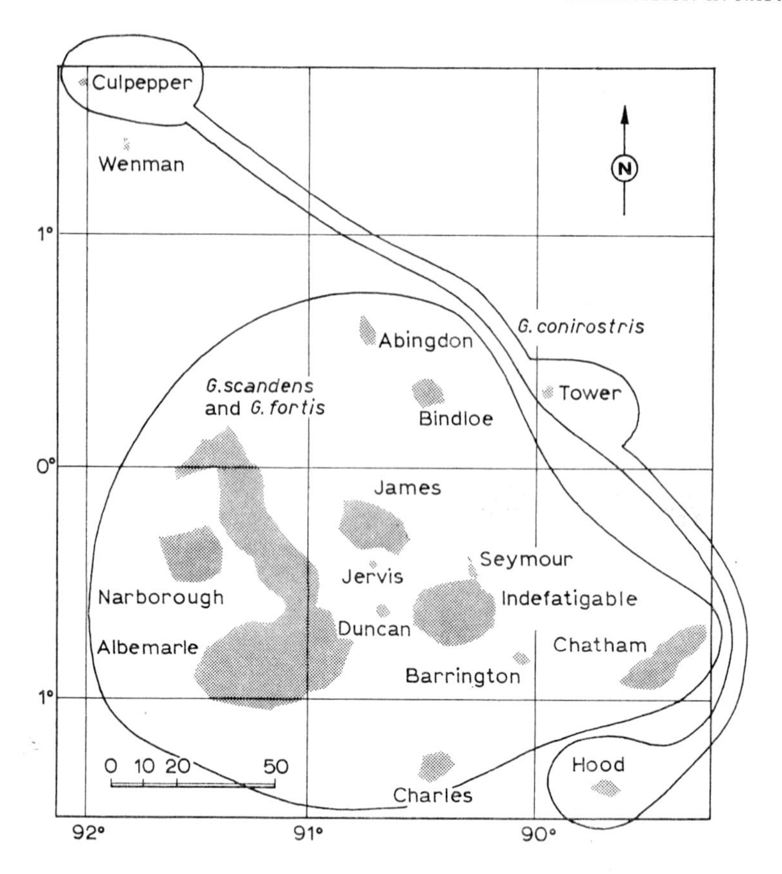

Fig. 12.2 The distribution of *Geospiza conirostris*, and the common distribution of *G. scandens* and *G. fortis* in the Galapagos Islands.

third. It has been discussed to some extent in Chapter 11. It is frequently found that two species are of different size when they occur in the same habitat, but in other places and in similar habitats where only one is found, the size is intermediate. In the British Isles there are two small mustelid carnivores, the stoat *Mustela erminea* and the weasel *Mustela nivalis*. Only the stoat is found in Ireland and the Isle of Man, while both are found in England, Scotland and Wales. This seems to be a case in which, after the last retreat of the glaciers, only one of the species spread sufficiently rapidly to get into Ireland, but the causes of the difference in distribution need not concern us here. The mean skull lengths are given in Table 12.1 (from Mayr 1963).

Table 12.1

Skull length, mm	Stoat G.B.	Stoat Ireland	Weasel G.B.
♂	50	46	39
♀	45	42	34

In cases like this it is certainly reasonable to suppose that the size difference in the skulls results from natural selection caused by competition. Even so the two species could still be having a detectable effect on each other's population: the evidence merely indicates without proof that an animal which is different in size from an animal of a closely related species is at a selective advantage.

Ecological differences in one habitat

The same sort of criticism can be levelled against the fifth category, the ecological difference of closely related species in the same habitat. It is nearly always possible to show what at first sight might be surprisingly large differences in some aspects of the ecology of species that occur together in the same habitat. It is interesting and informative for the study of the evolution of a group, but less satisfactory for ecological analysis. It is only possible to say that competition is smaller than might have been expected, not that it is absent or even that it is necessarily small. This category of evidence gives no information at all on the extent to which population densities are affected by the occurence of neighbouring species.

One example of this was given in Chapter 9, Broadhead and Wapshere's (1966) data on *Mesopsocus*. Here the major ecological difference between the two species was in the choice of egg laying sites. However as was shown in Chapter 9, the species are probably still competing to a very slight extent, that is to say affecting each other's numbers, because of their common parasites, even if interspecific competition for egg laying sites is absent.

Many examples of ecological differences in one habitat can be found in birds (Lack 1944), copepods (Hutchinson 1951) and flowering plants (Harper *et al.* 1961). Usually the largest ecological difference relates to what one might reasonably suppose to be the population controlling factors, and it would seem sensible to use this type of evidence heuristically to make preliminary decisions about what the controlling factors are. But only quantitative studies can show conclusively what the controlling factors are, or whether two species are or are not in competition.

Island and mainland populations

The sixth category of evidence compares populations on islands and on the mainland. Here the typical situation is that the same biomass per unit area is found in both places, but that there are fewer species on the island. So each species in the island has typically a greater average biomass than on the mainland and, again typically, it also has greater variance in its measurements (van Valen 1965). MacArthur and Wilson (1967, p. 97) describe the biomass effect as the best evidence of competition: to my mind it is very good evidence for the continuing stabilizing selective effect of competition, but is, if anything less informative about the extent of competition than categories 3, 4 and 5: ecological replacement, character divergence, and ecological differences, though it shares points with all these. Soule and Stewart (1970) have questioned this interpretation of the increase in the variance of measurements, and suggest three other possible origins for the increase, none related to competition.

Small islands in fact frequently have an impoverished fauna and flora compared with neighbouring larger areas. It is not really possible at the moment to say whether this is the result of competition under severe conditions, or just the absence of physically acceptable habitats for some species on small islands. Bristowe (1935) noted that very small islands frequently only have one species of spider from each of the common families, and in that sense the pauperisation is a systematic one. In another sense, this is what one would expect for a small random sample of the larger fauna on neighbouring islands (Greenwood 1968) and the systematic aspect is confined to the choice of which species in a family is represented. A good naturalist can usually predict fairly easily which species would be found.

With all this weight of evidence, it is not surprising that many ecologists, particularly terrestrial animal ecologists, regard competition as rare or non-existent in nature. The number of examples known in each of the categories discussed so far is now very large. I have thought it sufficient to give only the minimum of examples partly because they have been discussed so extensively in other books, but more particularly because only the first, displacement following invasion, can hope to produce much quantitative information on the strength of competition, even when it is transient. For a quantitative analysis of population processes, none of the others are satisfactory.

Together and apart

There are, though, two further categories of evidence which indicate that competition, quite reasonably strong competition, does occur in nature. One of these comes from the observation of distributions, the other, and more satisfactory one, from experiments.

The observational evidence is that some pairs of species occur together in some habitats, but only separately in others. The habitats in which they occur together are either clearly, or can reasonably supposed to be, the more favourable ones. The simplest interpretation of this sort of observation is that, under difficult conditions competition leads to the elimination of one or the other, but when things are easier stable competition is possible. Five examples seem to be reasonably satisfactory, though in each case there are still some unanswered questions. (Mayr (1963) gives further ones, but most of these depend on rather casual observations.)

The first example is of two species of lizard *Lacerta sicula* and *Lacerta melisellensis*. They occur together on the mainland of Yugoslavia, but separately and apparently at random, on some of the Dalmatian islands (Radovanović 1959). Here there is no evidence beyond that of distribution: the relation of the two species on the mainland is not known, and one is bound to be sceptical that the distribution on islands is a matter of chance until some detailed studies have been made.

The second example is of two wolf spiders *Lycosa* (or *Pardosa*) *monticola* and *L. tarsilis*. These belong to an interesting species group. In Britain there are five species in the *monticola* group, fairly clearly distinguished by habitat preference—salt marshes for one, pebble beds for another and so on. There are two forms in the *tarsalis* group. *Lycosa herbigrada* is the form that occurs in longish vegetation, and has a distinct colour pattern, but hybridises with *tarsalis* in the laboratory, and apparent hybrids have been found, though rarely, in the field. Locket and Millidge (1951) regard *herbigrada* as a subspecies of *tarsalis*. So *tarsalis* and *monticola*, the pair of the example, are not especially closely related, but they are nevertheless more closely related to each other than to any of the remaining 10 species of *Lycosa* on the British list. They are found in very short grass (1 cm long or less), and the colour patterns are almost identical, and make them cryptic at rest. Like all wolf spiders, they spin no web and catch their prey by running and jumping, and are easy to see when moving. On small islands off the west of Britain, *L. monticola* only is found on the southern ones up to Bardsey off North Wales. On the Scottish islands, only *L. tarsalis* is found. Their world distribution overlaps very widely. *Monticola* is found from Britain to asiatic Russia, *tarsalis* from Iceland through the palaearctic region to Kamchatka (Bristowe 1939, Williamson 1949). Bristowe (1935) says they are found together on the fairly large island of St. Mary's Scilly Islands. No quantitative study has been made of the microdistribution of these spiders in any areas where they are found together.

The third example deals for the first time in this Chapter with fresh water animals, and also the first quantitative data. Figure 12.3. shows the occurrence of five species of planarian in the British Lakes, (Reynoldson 1966). *Polycelis nigra* is the only one found in water with very low calcium concentrations. With rather more calcium, up to about 2.5 mg/l, *P. nigra* occurs only if *P. tenuis* is absent. Above 10 mg/l the species occur together quite commonly. These species are discussed again below.

The fourth example is, perhaps surprisingly, the only one with birds. Moreau (1948) made a very interesting study of the East African bird fauna to show to what extent the patterns of distribution matched those that Lack had found in British birds and Darwin's finches. He noted that in *Buccandon*, *B. olivaceum* and *B. leucotis* occur together in low and intermediate rain forests, but in two areas of high rain forest only one was found. *B. olivaceum* occurs in Poroto, which is in

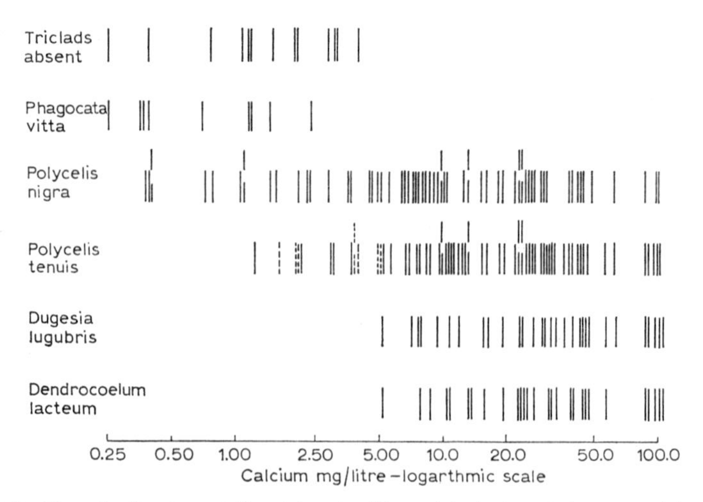

Fig. 12.3 The distribution of five lake-dwelling triclads in relation to calcium. Each vertical line represents one lake. Broken lines indicate the occurrence of *Polycelis tenuis* in lakes where *P. nigra* is not found. Note the exclusive distribution of these two species below 2.5 mg calcium per litre. (from Reynoldson 1966).

South West Tanganyika (Tanzania) near Lake Nyasa, *B. leucotis* in Usambara and Kilimanjaro in the north and east of the country. The distance between these two areas is about 500 miles and there is no other high rain forest in between.

The fifth example dates from Oldham's (1929) observations that on blown sand (but not on stabilized dune sand) round the coast of Great Britain, only *Cepaea nemoralis* or *Cepaea hortensis* is found, and that *Cepaea hortensis* is found only in this habitat north of Aberdeen, which is beyond the northern limit of *C. nemoralis*. The genetic polymorphism of these snails has been discussed in Chapter 7. Inland, most colonies consist of only one species, but quite a number of mixed colonies are known. However there have been casual reports of *C. hortensis* on English and Welsh dunes. It is certainly rare there and usually absent, though the distinction of dunes made by Oldham is sometimes difficult to draw. Oldham also made an interesting observation that two further species of snail, *Helix aspersa* and *Arianta arbustrorum* behave in the same way. Cain, Cameron and Parkin (1969) discuss the distribution of all these species on the north and west of Scotland, and emphasise correctly the difficulty in accepting even this type of observation as evidence for competition.

Helix is more closely related to Cepaea (which indeed is often considered to be a sub-genus of Helix) than it is to Arianta, but Helix and Arianta look more like each other than either looks like Cepaea. Cepaea has clear colours with well marked bands or none at all, Helix and Arianta are mottled brown with poorly defined bands. Helix and Arianta certainly occur together sometimes on sand dunes, but Helix is rather rare outside gardens except in south-west England.

The second pair suggests that some aspects of the appearance of the snails, possibly the impression they give to predators hunting by sight, is an important factor in preventing stable competition on the dunes. As was noted in Chapter 7, predators hunting by sight are still almost the only selective force known to be involved in the remarkable polymorphism in Cepaea. It would still be interesting to try to measure competition in a mixed inland colony.

Experimental evidence

There remains one category of evidence, and this is experimental evidence. If altering the number of one species, either up or down, alters those of its possible competitors, respectively down or up, and if the experiment also works the other way round, competition has been demonstrated. There are certain provisos, of course. These experiments must not alter the physical nature of the habitats, they should be done with controls, the numbers of other species must be held constant and any special explanations must be met. Field experiments are still far too rare, despite Varley's (1957) plea. A strictly theoretical experiment of this sort was described in Chapter 9, and the importance of allowing for changes in other species is shown there. There are in fact only two experiments with animals which seem to relate to competition in the sense that I am using it here. There are a number of other important experiments, such as Connell's (1961) that happen to relate to amensalism rather than to competition.

Reynoldson's work on planarians was mentioned on page 119, and an account of some other aspects of it will be found in Reynoldson and Davies (1970) and Reynoldson and Bellamy (1971). As part of his work he took *P. tenuis* from the 'College pond' at the University College of North Wales at Bangor. When he did, *P. nigra* increased. When he ceased taking *P. tenuis*, the two species returned to their former proportions. The figures are set out in Table 12.2. The unfortunate

Table 12.2 Number of *Polycelis tenuis* to each *P. nigra* in the College Pond, Bangor, before and after selective removal of *P. tenuis* in 1948–1952.

1948	1952	1954	1955	early 1956	mid 1956	late 1956	1957
4.85	0.43	0.49	0.79	1.46	3.80	5.00	3.67

(Data from Reynoldson 1966)

point on this experiment is that it still does not show if *P. nigra* effects *P. tenuis*. The only figure that helps is the rate of return to the old density of *P. tenuis*, and there are no comparable figures for this in the absence of *P. nigra*.

The other experiment with animals is also inconclusive, and the animals are plant like in some important ways. Pontin (1961, 1963, 1969) studied two common species of British ant, *Lasius flavus* and *L. niger* on Bowling Green Alley, near Oxford. The ants live in colonies that are, like plants, established at one particular place. Pontin found that:

1 Queen production is a good measurement of colony size.
2 For *L. flavus*, but not *L. niger*, the distance apart of colonies is proportional to the square root of the number of queens. This is a statistical demonstration that the colonies fill the available space, and confirms casual observation.
3 When *L. niger* colonies are also present, the queen production of *L. flavus* is depressed to about a quarter of its expected size.
4 Removal of *L. niger* colonies by poisoning leads to
   (i) an increase in neighbouring *L. niger* colonies (implying intraspecific competition)
   (ii) an increase in *L. flavus* next to the *L. niger* colony that had been, destroyed, but
   (iii) a decrease in *L. flavus* on the far side of the remaining *L. niger* colonies. Altogether this shows that *L. niger* depresses the number of *L. flavus*
5 Placing *L. flavus* nests (by digging them up) round *L. niger* nests reduces queen production in *L. niger*, but also stops queen production in *L. flavus* completely.

The interesting thing about these two species is that they are quite clearly distinct in many ways, although at the same time very similar. *L. flavus* is entirely subterranean, and forms compact three-dimensional nests about 1 m across. *L. niger* is much more a surface liver, and has more diffuse nests which are about 4 m across, and only about a third of the workers forage underground. Both species are largely aphid eaters. There are no less than fourteen species of aphids fed on by both, with no evidence that the ants differ in their preferences.

Pontin also points out that his colonies are probably still the immediate descendents of the first queens on the site, and so an historical singularity is involved. It could be argued that this is an experiment on individual competition, rather than a populational competition, and that a complete system requires new colonies to be established. On the other hand there have been many generations of workers in the colonies at the site.

Those two series of experiments were on animal populations. Some rather differently designed experiments on plant populations have been done by Harper and his colleagues. These involve adding selective herbicides to swards and following the changes thereafter. The interpretation of these experiments is appreciably more difficult than the interpretation of either Reynoldson's or Pontin's, but when the method has been fully developed it will probably be of more general use. The difficulties come to some extent because every herbicide damages more than one species, but also because of the problem of distinguishing between changes from competition and changes from the reduced density of the vegetation. In addition to removing plants by herbicides, others may be added, usually most conveniently as seeds. One example will show the techniques and the difficulties.

Putwain, Machin and Harper (1968) studied the regulation of a population of *Rumex acetosella* in a community of more than eleven other species of plants, which included six species of grass, and five other species of dicotyledonous herbs. Experiments in which seeds were sown, but herbicides were not used showed that the population was maintained almost exclusively by vegetative reproduction. The method of analysis that was used was the approximation to the Moran method mentioned in Chapter 5. In a later experiment (Putwain and Harper 1970), one treatment removed all the grasses, another the herbs except Rumex, a third removed all plants except Rumex, and the fourth treatment removed only the growing plants of Rumex. All three components, Rumex, grasses and the herbs, gained by the removal of any of the other classes, and similarly Rumex seedlings could establish themselves when any one of the classes was removed.

It seems that all the species in the sward may be in competition. So far the interpretation has been by speculative diagrams of 'fundamental niches', which give an indication of the sort of mathematical model that might soon be developed. Unlike the other cases in this Chapter, the competing species are not particularly closely related, and this problem leads on to those of Chapters 15 and 16, to systems of many species.

The evidence for stable competition in natural populations is extremely tenuous, but at the moment it looks as if more experiments might well show that mild competition is quite common. Heuristically, the study of competition may often indicate what are the controlling factors of the species, and so speed the study of the population dynamics of the individual species. Most studies so far have told us more about evolution than about population dynamics; more about genetics than ecology; but field experiments offer one of the best hopes of combining all three into an integrated population analysis.

13 Feeding interactions in nature

Of the five classes of interaction between two species which were defined in Chapter 9, only two give rise to major problems in analysis. These are the negative-negative interactions, discussed in Chapters 10, 11 and 12, and the positive-negative ones that will be discussed in this Chapter and the next two. This second class involves primarily the feeding of one species on another. Judged by the population effects, it is likely that feeding sometimes produces other classes, such as +0, when the population consumed is not affected, or 0− when the consumer population fails to gain in numbers. How frequently this happens is not known, for the most striking fact in the analysis of these systems is that hardly anything is known about the strength of the interactions from the point of view of population dynamics. There is an important zone of contact here between behaviour and ecology. Quite a lot is known of the quantitative aspects of feeding behaviour, for instance Ivlev (1960) or fishes, or Gibb and Betts (1963) on feeding by birds, to mention only two of the more ecological studies, but these have not on the whole been extended to become population models. An exception is the work on marine plankton, discussed below.

There is a strange contrast between feeding interactions and competition. It is quite clear when one organism feeds on another, it is difficult to show that one is in competition with another: but competition experiments are easy to set up in the laboratory, while predator/prey experiments, which will be discussed in the next Chapter, have mostly been unsatisfactory. One possible reason for this is suggested by the variation in specificity of different types of feeding interaction.

Classification and specificity of feeding interactions

All living organisms can be divided into autotrophs or heterotrophs. Autotrophs use only inorganic compounds, and use either the energy of sunlight (phytotrophs) or, much more rarely, chemical energy (chemotrophs) to build up complex organic compounds. Heterotrophs, therefore include almost all organisms except green plants. As usual with these sort of classifications, there are difficult cases. All lichens and quite a few animals, such as *Paramecium bursaria*, and some corals, are mixtures of two species, one an autotroph and one a heterotroph. Heterotrophs may be grouped into those that require living food, and those that feed on dead material, including saphrophytes, but these two groups grade into each other. For instance amongst the birds of prey, one can find a range of species from falcons which catch other birds in flight, to vultures which feed only on carrion. The classification given below is for heterotrophs feeding on other living organisms. Two criteria are used in this classification. The first is whether the consumer kills and eats the whole individual it is attacking, or only consumes part of it. The second is whether the consumer attacks lots of individuals in its life time, or whether the life cycle can be completed within one individual. These two criteria give four classes.

1 Consumers killing lots of whole organisms include most normal predators, especially the large ones, and certain herbivores. Examples are zooplankton feeding on phytoplankton, whales, robber flies, spiders, swallows and other insectivorous birds, stoats and weasels, planarians, and fish like trout. The range of structure and habits is very wide, but typically this class of consumers attacks a wide range of species, and the specificity of their attack is determined by the size of the prey, the place and time at which prey are found, and to some extent by preferences of taste, and not by specific differences in their prey. For instance Oldroyd (1969) quotes the case of the Asilid or robber fly *Promachus negligens* which has a diet known to include five species of moths, five ants, one ichneumon, four wasps, two bees, five beetles, one dragon fly, one ant-lion, one termite, two horse-flies and several species of small fly.

2 The consumer takes part only of several individuals. Most terrestrial herbivores of any size come here, as is noted by Harper (1968). Again these tend not to be species specific. This class also includes organisms such as bloodsuckers, like fleas, ticks, mosquitos and horse flies. Some of these, particularly fleas, are exceedingly species specific. Even if they can feed on the wrong species, they cannot normally reproduce on it.

3 The consumer uses all of one whole organism to complete its life history. The important group here are the insect parasitoids. In these the female lays an egg in a larva of another insect, and the whole of the life until the emergence of the adult is spent within that individual that has been attacked. The adult may require some other food. In this group, high specificity is quite common but it is also fairly common to find species that will attack a wide range of prey species. There is an interesting discussion of the effect of this in Zwölfer (1971), which will be mentioned again in Chapter 17.

4 The final class is that in which parts of only one organism are attacked. The typical organisms here are pathogens and parasites, and again species specificity is common.

This classification emphasises the wide range of feeding interactions, and so the difficulty of finding any common population pattern amongst them. Entomologists have paid a lot of attention to class 3, but other ecologists tend to think of class 1 when thinking of feeding interactions, and indeed frequently refer to the whole set as predator/prey interactions. It is worth noting that this is the class where specificity is lowest, and therefore the class in which analysis is likely to be most difficult. The complexity of the subject suggests that a matrix notation might be helpful, but this appears not to have been used. For instance it might be possible to construct transition matrices converting the frequency of food organisms in the environment to the frequency of food organisms captured by particular species. Probably not enough is known at the moment to say what the functional form of the elements of such a matrix should be.

Other aspects of feeding behaviour may well be important in analysing these interactions in nature. For instance the method of capture may be important. Certain organisms lie in wait for their prey, notably sessile aquatic predators and spiders, while some of those that run after their prey hunt individually, some in packs. Similarly the prey show varying degrees of non-randomness from scattered individuals to closely aggregated shoals.

Predator/prey interactions

The feeding interactions usually studied by ecologists, as opposed to parasitologists or epidemiologists, are often predator/prey interactions. An interesting attempt to make generalizations in this field is found in Huffaker (1971). Besides the point already emphasized that these interactions are frequently not specific, certain other features are also commonly seen. Predators not infrequently seem to find it difficult to catch enough prey. They may spend much time looking for it, and they may frequently miss when they attack. This suggests a fine balance of advantages developed during the course of evolution, and this will be discussed more in Chapter 15. There has been much dispute as to whether predators only take certain limited classes of the population they are attacking such as those out of cover, and those who have not established territories, and those not in full health. No doubt such prey are easier to capture, but restrictions to these classes is probably not particularly common.

There are rather few data on the relative changes in the numbers of a predator and its prey. It seems fairly certain that the numbers of lynx in Canada (Fig. 1.7) follow the numbers of their prey, the snowshoe rabbit, *Lepus americanus*. In weekly records of plankton in the north-west North Sea, the numbers of the gymnosomatous pteropod *Clione limacina* appear to follow those of its prey, the thecosomatous pteropod *Spriatella retroversa* (unpublished data of the Scottish Marine Biological Association, Edinburgh. See Williamson 1961a). The predator/prey system that possibly has been most discussed is another one in plankton, namely the feeding of zooplankton on phytoplankton. Both of these groups contain many species, but interesting and sensible analyses can be made considering them as groups, thereby refuting the common dogma that no ecology can be done where individuals are not identified to species. Figure 13.1 shows the variation in the plankton seasonal cycle in the sea around the British Isles. Four patterns have been distinguished here in the phytoplankton, differing mostly in the size of the spring and autumn peaks, while there seems to be a common seasonal cycle for copepods in all areas. Similar cycles occur in fresh water lakes. There can be no doubt that the summer increase of copepods depends on the spring increase of phytoplankton. The main point that has been disputed is

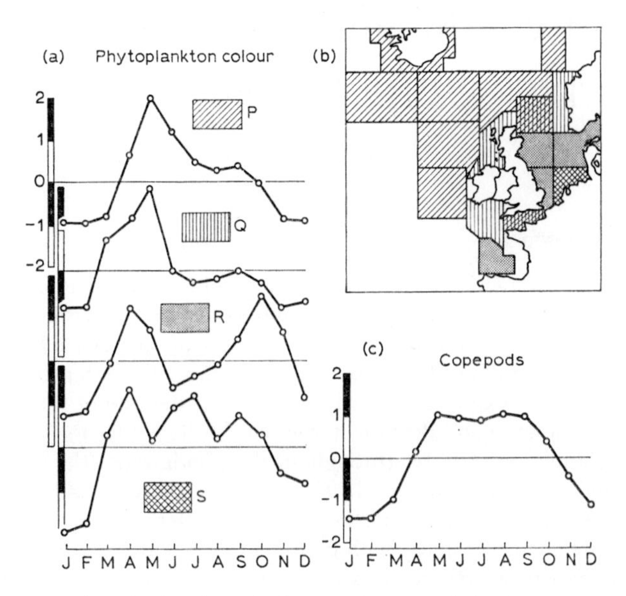

Fig. 13.1 Seasonal cycles in phytoplankton and copepods in the sea round the British Isles, from Colebrook and Robinson (1965). The graphs marked a show the standardised mean seasonal variation in abundance of phytoplankton in the groups of areas indicated by different hatchings in the chart b. c shows the mean seasonal variation of copepods in all areas. Details of phytoplankton cycles across the Atlantic are given in Robinson (1970).

whether the summer decline of phytoplankton is caused by grazing by zooplankton, or whether it is the result of death from nutrient depletion. Lund (1965) emphasizes the importance of nutrient depletion, Cushing (1964) the importance of grazing. Riley (1963) suggests reasonably that both will frequently be important and it is also reasonable to suppose that their relative importance will vary in different communities. This interaction has lead to some interesting theoretical models which can be considered along with some experimental studies on predator/prey interactions in the next Chapter.

14 Experiments and the theory of predator/prey interactions

Grazing in the sea

The theory of interactions between phytoplankton and zooplankton must include a number of constituent parts, for some of which it is difficult to build a simple realistic model. Consider first the physics of the sea. The energy for growth comes from sunlight falling on the sea surface. The intensity of the light declines more or less exponentially with depth, but the rate of this decline will depend on the plankton and detritus in the sea. In the winter in most temperate areas there is a mixing of water throughout the photic zone (the zone in which there is enough light for growth). In the summer, many areas have an upper, mixed warm layer, separated from a lower, mixed cooler layer by a zone in which the temperature changes rapidly with depth, a thermocline. For making models, the summer sea can be regarded as two homogenous layers with some slight mixing between them.

The total phytoplankton can be measured in a number of ways; by the total chlorophyll; the total carbon content; or even by the total cell count, though the large differences in the size of the different species make the last one unsatisfactory. Most algae are heavier than sea water, and so will sink in still water, and although this will be counteracted by turbulance, there will still be a net loss from the upper to the lower layer in a two-layered sea. There will also be a loss of algal cells from grazing by zooplankton and from attacks by parasites, and a loss of algal weight from respiration. All these losses must be counterbalanced by gains by photosynthesis. The rate of photosynthesis will depend on the light intensity, the temperature, and possibly the concentration of certain nutrients such as phosphate, nitrate or vitamin B_{12}. The importance of these nutrients has been disputed and as Eppley and Strickland (1966) say 'the relationship between nutrient concentration and growth . . . has not been well defined, and in fact may not exist as such.'

There are two fairly well developed sets of equations for the dynamics of marine plankton, those of Riley and those of Steele, summarized and compared by Riley (1963). Both these assume that phosphate is a nutrient that affects phytoplankton growth, and both have arbitrary, logistic, functions for the growth of zooplankton. Steele's equations can be written

$$dn/dt = m(0.70 - n) - p(0.58n - 0.027)$$
$$d\ln p/dt = 0.75n - 0.11 - m - 0.024h$$
$$dh/dt = 4p - 0.01\,h^2$$

where n is phosphate in $\mu g.at./l$; p phytoplankton and h the herbivores, both in g carbon $/m^3$ and m the idealised coefficient of vertical mixing, a pure number. The constant in the middle equation combines respiration and sinking. Other equations have explicit expressions for the relation of phytoplankton growth to light and its absorbance. These equations may be compared with those for continuous cultures in Chapter 4. In these equations some of their terms, such as the last one in h^2, are arbitrary, but on the whole they represent, as do the terms in the continuous culture equations, distinct measurable processes. Respiration comes into the yield constant in continuous cultures, and the form used for respiration in the phytoplankton equations can obviously only apply over a certain range of phytoplankton densities. The major difference is that the plankton equations are intended primarily for short term prediction, particularly of the transition from the spring blooming of phytoplankton to the summer situation of much zooplankton and relatively few algae.

These equations have another feature in common with the continuous culture equations and, indeed, with any set of equations describing a feeding interrelationship. This feature is the double conclusion that if there is more food this will lead to an increase in consumers, and, conversely, if there are more consumers, there will be less food. In continuous cultures, an increase in the substrate concentration in the reservoir leads to a higher microbial concentration in the culture, while an increase in the dilution rate leads to a lower microbial concentration and to a higher concentration of the limiting substrate in the culture.

Predators on land

In considering the sea and continuous cultures, it is reasonable to assume at first that all or part of the system, such as the upper layer of the sea, is homogeneously mixed; that interactions between the components of the system are simple functions of their concentrations; and that no special allowance is needed for any delay from reproduction. When dealing with arthropod or vertebrate predators on land, all three of these assumptions are suspect, so the analysis of such systems is more difficult, and it is not surprising that the models for them are even less complete. Holling's work on the components of predation, summarized in Southwood (1966) and Holling (1968), has attracted a lot of attention, but has not yet lead to a complete analysis of a predator/prey system. As Holling says, there are a lot of details that need precise analysis in such a complex system. It is naive to think that generalized models will be adequate, even though all models may well have general features in common.

One important aspect of the analysis of these complex predator/prey systems is the study of the response of the predators rate of feeding to a change in the density of the prey. A particularly satisfactory analysis of this part of the system is Hassell's (1966), which is discussed in relation to the 'check-method' of evaluating the effect of predators by Huffaker *et al.* (1968). The 'check-method' is essentially the experimental removal or inhibition of an enemy in one environment so that its importance in another environment may be evaluated, and furnishes a very convincing demonstration that predators do affect the number of their prey. As with Putwain and Harper's (1970) use of herbicides, discussed in Chapter 12, the quantitative analysis of 'check-method' experiments is difficult.

In classifying the response of a predator population to a change in a prey population, Hassell distinguishes between the behavioural response and the intergenerational response. The behavioural response is observed and measured within one generation of the predator. Its importance is that the prey will usually have an uneven distribution in space, and so by the behavioural response, the predator population becomes related to the different prey densities. If the percentage of predation is higher in regions of higher prey density, i.e. directly density dependent, Hassell terms it 'superproportional', while if the percentage predation falls with increasing prey density, he terms it 'subproportional'. These relationships arise partly from a component of individual predator behaviour, and partly from the aggregative response of the predator population. Affects of satiety, of reduced time for searching, and also of immigration and emigration are involved, and as Hassell points out, could lead to a change from subproportional to superproportional response at different densities.

With insect parasites in particular there is likely to be an important intergeneration effect. This is because the generations of parasitoid and host are each distinct and, typically, the same length as each other. In other feeding interactions, an overlap of generations is frequently found, leading to a more continuous response of predator population to prey and vice versa. With insects, an increase in the number of host larvae available will lead to an increase in the number of parasitoid females ready to attack only one generation later. The increased numbers of females have to develop inside the host larvae. There can of course be a direct behavioural dependent response (Hassell 1971) but the population increase, the intergenerational response, must inevitably be delayed. As in other feedback systems, the delay could lead to oscillations, in this case in the numbers of both parasitoids and hosts.

Models of insect parasitoid systems

Several mathematical models have been built to describe the interaction between a parasitoid and a host, of which a classic one is that by Nicholson and Bailey (1935). Parts of this model were used by Broadhead and Wapshere (1966) in their model which was discussed in Chapter 9. A recent modification has been suggested by Hassell and Varley (1969) who test their ideas with several sets of data, and explore the theoretical consequences. The critical part of the model is the proportion of larvae attacked at different parasitoid densities. If u_i is the initial host density and u_f the final host density, the number of parasitoids developed is $u_i - u_f$. If the density of adult parasitoids attacking u_i hosts is p, Hassell and Varley suggest that

$$ln\ u_i - ln\ u_s - Qp^{(1-m)},$$

where Q and m are constants, the quest constant and the mutual interference constant. If there is no mutual interference amongst the parasitoids, m becomes zero and the model is the same as Nicholson and Bailey's, in which there is only one constant 'the area of discovery', defined as $a=Q/p$. A disadvantage of the earlier model is that it leads to increasing oscillations. In the new model, with a Q of 0.1 and the host power of increase of 5 then stable oscillations are produced at $m=0.5$, with damped oscillations for larger m, though the equilibrium reached is, somewhat surprisingly, at the peak of the stable oscillations and not at their mid-point. With this model, several parasitoids can coexist when attacking the same host.

This model has important implications for biological control. Hassell and Varley conclude that a pest would be best controlled by releasing as many of its parasitoids as possible. Several entomologists dispute this and claim that the best strategy is to introduce only the most effective parasitoids. In support of this latter view, one can quote Watt (1964, 1965) who found in the data of the Canadian forest insect survey, that increasing the number of attacking species decreased the stability of the species attacked. This will be considered again in Chapter 17. Zwölfer (1971) gives evidence that the specialized, stenophagous, species which would make good biological control agents, are often at a competitive disadvantage to the less effective unspecialized polyphagous species. This consideration has not yet been tested by Hassell and Varley, and further developments of their model should be interesting.

General models and microbial experiments

Hassell and Varley's model predicts the numbers at all stages of the life history of both parasitoid and host. It is a model for the complete generation cycle. It includes an expression for the rate of attack of one species on another, but certain other models, such as Watt (1959) or Rosenzweig and MacArthur (1963) are, explicitly or implicitly, models only of the rate of attack. For a generalized, continuously reproducing predator there seems to be only one reasonably satisfactory model, that of Leslie (1948). There are other specialized models which would take too long to discuss here, among them the mathematical theory of epidemics, which is one of the few areas in which stochastic models have been developed to advantage in population ecology (Bailey 1957, 1967; Bartlett 1960).

The rational basis of Leslie's model can be seen by considering, as in the theory of competition in Chapter 11, the loci of equilibrium separately for predator and prey. As was noted when discussing the equations for marine plankton, more prey would mean more predators, so the equilibrium locus for predators should have a positive slope, while more predators would mean fewer prey, and so the equilibrium locus for prey should have a negative slope. Writing H for the prey (assumed to be a herbivore) and P for the predator, Leslie's equations are

$$d \ln H/dt = a_1 - b_1H - c_1P$$
$$d \ln P/dt = a_2 - b_2P/H$$

The loci for these are shown in Fig. 14.1a, and it can be seen that they imply an oscillation which Leslie showed was a damped one.

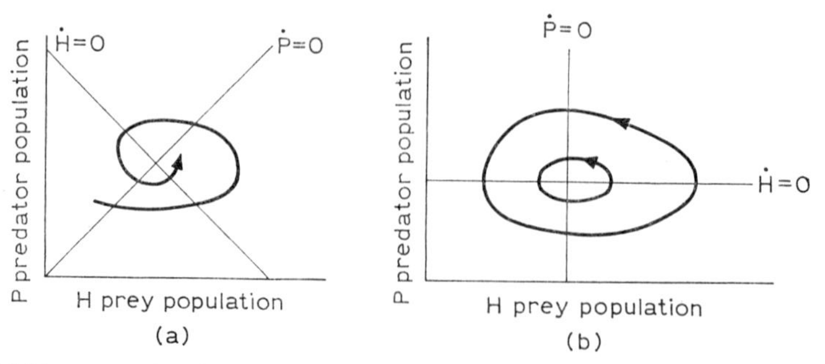

Fig. 14.1 Diagrams of predator/prey theories. Each diagram shows the predator population (P) plotted against the prey (or herbivore) population (H). The loci of $d \ln P/dt = 0$ and $d \ln H/dt = 0$ are marked $\dot{P} = O$, and $\dot{H} = O$, and the expected course of the pair of populations is shown. The left hand diagram, a, shows Leslie's (1948) system leading to a stable equilibrium point by damped oscillations. The right hand diagram, b, shows Lotka's (1925) simple system, leading to constant metastable cycles.

These equations can be contrasted with Lotka's and Volterra's

$$d \ln H/dt = a_1 - c_1P$$
$$d \ln P/dt = c_2H - b_2$$

whose loci are shown in Fig. 14.1b, and are clearly unacceptable. The prey are in equilibrium at any density provided there are exactly the right number of predators, and similarly for the predators. These classical equations lead to closed cycles of the two species, so that the size of the oscillation would appear to depend on the initial densities. However this is one case where the continuous approximation to what must, in any real case, be a path across a rectangular grid of points is unsatisfactory. Because the path must go to a grid point at each step, a little calculation shows that the paths of Fig. 14.1b are impossible as a generality, and the population could show either damped oscillation, or increasing oscillations, or possibly just converge to one particular limiting cycle. These considerations are not important here, because of the nature of the loci, and it can be said that any model producing similar loci is also unbiological.

Leslie's equations are a simple modification of the logistic equations, and must therefore also be regarded as rough first approximations. The equation for the prey is the logistic equation with the addition of a linear term for mortality from the predator. The equation for the predator is a logistic equation in which the density dependence of the second term is modified to take account of the number of prey.

In addition to his experiments with competition, Gause (1934) also tried experiments with predator/prey systems, but was unable to prevent one or other being eliminated without some rather artificial manipulations of his cultures, such as regular reinoculations. Leslie and Gower (1960) have re-examined one of Gause's best known experiments in which *Paramecium caudatum* was the prey and another ciliate, the gymnostome *Didinium nasutum* was the predator. Didinium attacks Paramecium one at a time, but can consume several fairly quickly. Leslie's (1948) equations were used, modified for calculation in discrete steps,* and the system was simulated on a computer, with the parameters subject to some random variation. They found three points which need mentioning here.

The first is that the system scarcely showed any cycles when the parameters were adjusted so that the equilibrium was 150 Paramecium to 100 Didinium. The result of such a computer run is shown in Fig. 14.2. Around time 100 there

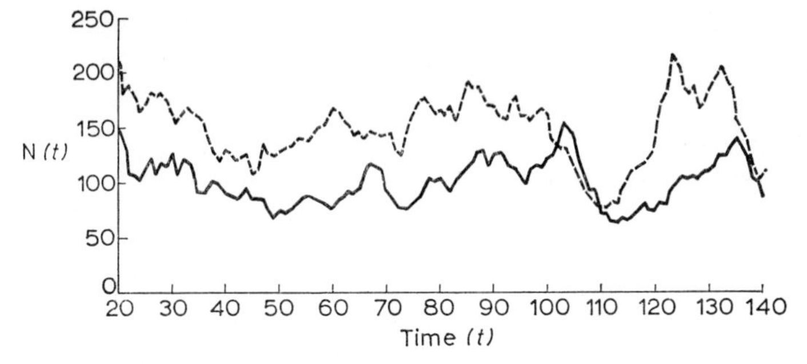

Fig. 14.2 Results of a computer simulation of a predator-prey system, by Leslie and Gower (1960). The number of prey ('Paramecium') are shown by the dashed line, the number of predators ('Didinium') by the continuous line. The equilibrium point is150 prey and 100 predators.

is a small section of an obvious oscillation. Because of the relation of the equilibrium loci, it might be thought that the populations would be uncorrelated. In fact a small positive correlation is expected, and a correlation $+0.17$ was observed.

The second point is that Gause worked with too small a volume for the system to have a reasonable probability of persisting. The equilibrium expected in Gause's experiment was only five of each species, so it is not at all surprising that either the predator or the prey (and then necessarily the predator) became extinct in each replicate. Table 14.1 shows the results obtained by Leslie and Gower. It looks as though cultures only four times as large would have been satisfactory, while for the systems shown in Fig. 14.2 the expected time to random extinction of one or other species is 1.4×10^{11} units, or in other words never for practical purposes.

* Each step is one third of the division time for Paramecium and one quarter of the division time for Didinium.

Table 14.1 Results of Leslie and Gower's (1960) simulations

Paramecium at equilibrium	Didinium at equilibrium	Extinctions	Computer run
150	100	nil	1
23	15	399 steps, Didinium almost extinct once	1
18	9	Didinium 5 times Paramecium once	5
5	5	Didinium 10 Paramecium 9	19

The third point is that even on the small scale that Gause worked on, some protection for Paramecium which would render a proportion of the population immune to attacks by Didinium would have allowed the system to persist. This is shown in Table 14.2. It can be seen that even when the expected populations

Table 14.2

Proportion of attackable Paramecium	Paramecium equilibrium population	Didinium equilibrium population	Order of magnitude of the mean passage time to zero in units Paramecium	Didinium
0	250.0	0	5.8×10^{40}	
0.05	230.1	11.5	9.8×10^{34}	6.2×10^{2}
0.10	194.0	19.4	6.0×10^{25}	1.3×10^{4}
0.15	155.7	23.4	7.4×10^{17}	4.8×10^{4}
0.20	125.0	25.0	1.2×10^{13}	6.5×10^{4}
0.25	90.7	22.7	1.5×10^{8}	1.7×10^{4}
0.35	50.4	17.6	2.4×10^{4}	1.5×10^{3}
0.50	23.3	11.6	3.3×10^{2}	1.6×10^{2}
0.65	12.8	8.3	76	58
0.80	8.1	6.5	37	34
1.00	4.9	4.9	20	21

are only five and five, if all the Paramecia can be attacked, if 65% to 90% of the Paramecia are protected, the system is reasonably stable.

A continuous culture experiment

Another possible way of stabilising a predatory/prey system is to use a continuous culture. There are considerable technical difficulties but the loss to both populations by the continual dilution should damp down oscillations fairly quickly. The only examples of this known to me are some of my own unpublished work. The results of one such experiment are shown in Fig. 14.3. The prey is a yeast *Schizosaccharomyces pombe* growing on a defined medium, and the predator another ciliate *Tetrahymena pyriformis* which cannot grow on that medium. The culture was not completely stable, but it can be seen that it

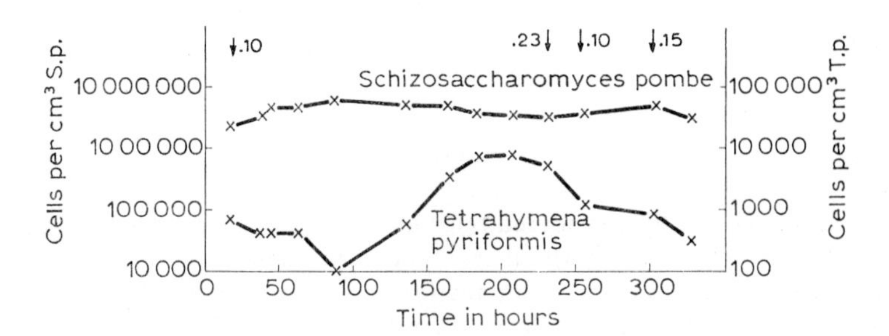

Fig. 14.3 A predator-prey system in continuous culture. The left hand ordinate gives the number of prey, the yeast *Schizosaccharomyces pombe,* per ml, the right hand ordinate the numbers of the predator, the ciliate *Tetrahymena pyriformis.* The figures across the top indicate the start of particular dilution rates. (Mark Williamson, unpublished data.)

persisted satisfactorily for a while, that the Tetrahymena did vary rather more than the pombe, and that an increase in Tetrahymena lead to a decrease in pombe. Tetrahymena is about 100 times the volume of pombe, and so the scales have been positioned so that the volumes are more or less comparable.

All these studies come to much the same general conclusion. It is possible to analyse predator/prey systems. Realistic equations will probably imply damped oscillations. Even very simple systems can be stable and persist. Considering the importance of these systems, it is remarkable that so little is known of their ecology, and as will be seen in the next Chapter, even less is known of their genetics.

15 Genetical consequences of two species interactions

Geneticists have mostly considered selection acting on a single population, and have not attempted to study the more difficult problem of selective interaction between two species. Yet these interactions are clearly shown by evolutionary studies to be important. At the level of speciation, the influence of competition in separating two populations has been much argued, and what quantitative analytic work that has been done on this is mentioned in Chapter 11. At higher levels, adaptive radiation has frequently involved the development and exploitation of new food sources, and the coevolution of the predator and prey. Much of the evolution of flower structure in the angiosperms must be associated with pollen and nectar eating insects, and insects have evolved to match the flower structures. It is reasonable to suppose that much of the diversity both of the consuming species and those consumed comes from coevolution. This has been urged for butterflies and plants by Ehrlich and Raven (1964), suggested for freshwater plankton by Lund (1965), and Harper (1968, 1969) speculates on the importance of herbivores in the ecology of land plants.

There is rather little in the theory of predator/prey interactions that bears on genetics, and I know of no experiments at all in this field. The main relevant point in the theory is Volterra's third law, which is given in English in D'Ancona (1954) and Kostitzin (1939). Volterra's three laws for predator/prey systems are:

1 the law of the periodic cycle
2 the law of conservation of averages
3 the law of the disturbance of averages

The first two state that predator/prey systems will tend to cycle, and that the average of each population over a whole cycle will not drift. If damped oscillations are allowed, these laws are reflected in all the formulations of predator/prey systems considered in the last Chapter. The third law is the one of interest here. This states that if the mortality of both predator and prey are increased simultaneously, by some other factor acting on both, then the number of prey will increase, but those of the predator will decrease.

This law could have odd genetical effects. Consider a strictly hypothetical example. Suppose that populations of the snail *Cepaea nemoralis* in beech woods are controlled by an insect parasite, so that the snail/insect system can oscillate. Then if thrushes attacked parasitized and non-parasitized snails at random, and even more if the thrushes attacked the parasitized snails preferentially, then an increase in thrush predation would lead to an increase in the snail population and not to a decrease. Now, as was shown in Chapter 7, thrush predation is often related to the degree to which the snails match their background. So a population of Cepaea which were all phenotypically yellow, and which would be conspicuous, would under these assumptions become more numerous than a population of cryptic brown Cepaea. This example shows another theoretical difficulty in the concept of population fitness. Under certain conditions the population that is better adapted will, because of this adaptation, have a lower equilibrium population size. This example is artificial in at least two ways. The factors controlling the population size of the Cepaea are not known, and might well act in the egg and recently hatched period of the life history rather than the adult. The evidence on the thrush predation refers to the relative effectiveness of thrushes against phenotypes in one population, while the example assumes that this would carry over to two populations each composed of one phenotype. I have chosen to use it here both because there seems to have been no attempt to study real examples of this phenomenon, and because of the discussion of the polymorphism of Cepaea in Chapter 7.

Volterra's first and second laws seem to apply in most models, but does his third law? The answer seems to be not always completely. Considered the effects on Leslie's model. Extra random mortality there is mathematically equivalent to a decrease in the intrinsic rate of increase. Some numerical manipulations of the parameters of the model given in Leslie and Gower (1960) will show what could happen, and these results are easily shown by algebra to be general results in this model.

The values used for the parameters were

$$a_1=0.2574 \qquad a_2=0.1892$$
$$b_1=0.0005148 \qquad b_2=0.2838 \qquad c_1=0.0018018,$$

in which the a_i are the intrinsic rates of natural increase, which gives an equilibrium for species one (Paramecium) of 150, and of species two (Didinium) of 100. A random mortality rate of 0.05 was used first on Paramecium, then on Didinium, then on both. In the first case, both are reduced, Paramecium to 81%, and Didinium to 81%. Mortality on Didinium again reduces it, but increase the number of Paramecium to 123%, the Didinium go to 90%. Both these results are quite clear from the algebra and consequently it is to be expected that attacking both will reduce Didinium, but the two effects on Paramecium seem, algebraically, almost to cancel. The result is Didinium reduced to 73%, but Paramecium only to 99% of its original population. In other words, Volterra's third law does not hold for the prey species in this system. From an evolutionary point of view, predators will be selected if they can avoid being involved in extra mortality on either themselves or their prey, though the latter is difficult to envisage. It is perhaps even harder to see how the prey could gain from a selective force that increases the mortality of the predator. No doubt examples could be found, but the results do not appear to lead to any general theory of coevolution.

A more promising variant is to change the parameter c_1. Reducing this is giving the prey a better chance of escaping the predator, or conversely reducing the efficiency of the predator. Doing this, perhaps surprisingly, increases both species in the same proportion. Another possible change in the efficiency of the predator is to change b_2, though this perhaps suggests a change in metabolic efficiency rather than in efficiency of pursuit. The effect is certainly different: predators increase and prey decreases if b_2 is decreased. Once again it seems there is a possibility that the predators that search out their prey more efficiently, which certainly could be a selective advantage, will have a smaller equilibrium population size. With predators, there is certainly no reason to suppose that the equilibrium population size, or the intrinsic rate of natural increase will be easily related to the fitness of the individuals, and so there is no justification for using either of these as a measure of that obscure quantity, the fitness of the population.

All these results are very tentative, and based on very simple models. The common occurence of predator/prey interactions, and the coevolution of these systems, suggests that they would repay more detailed study by population geneticists.

IV Many species

16 Multivariate descriptions

Any sample or survey of a habitat will, with a few extreme exceptions, show far more than two species living there: normally there are many, of many different sizes and ways of life. These are usually referred to as a community, which Fager (1963) defined as 'a group of species which are often found living together'. For one place, this means that the species normally found there are part of a community, and that casual occurrences of other species are ignored. Because of the different sorts of sampling needed, and because of taxonomic problems, only part of a community can usually be studied in any detail. An example of such a part community is the plankton in the north-west North Sea shown in Fig. 16.1. The method of sampling, by a high speed net, excluded species which passed through the net such as the μ-algae, or were too large to enter the net, such as adult fish or were too badly damaged to be recognizable as such, such as Ctenophores.

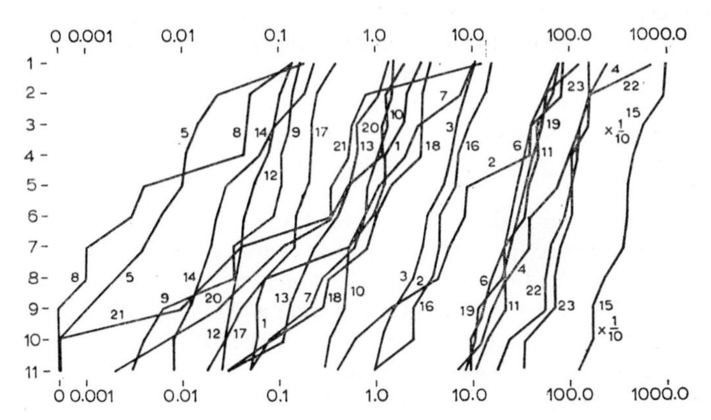

Fig. 16.1 The abundance of twenty-three entities (species, age classes of a species, or species groups) of plankton per third of a cubic metre per summer in the north-west North Sea. The figures for each entity are arranged in rank order, independently of the other entities, from highest to lowest (from Williamson 1961a).

Nevertheless the data form an appreciable sub-set of the community, covering seven orders of magnitude in numerical abundance.

Description by trophic layers

There are two usual ways of making the data on a community of many species more manageable and more intelligible. The first is to deal only with one taxonomic group, such as birds or higher plants. The second is to classify each species by its source of food, into its 'trophic layer', into primary producers, herbivores, primary carnivores, decomposers and so on. Both approaches have produced important results in community ecology, but neither seems particularly well designed to extend the analysis of population change from one or two species to many.

The trophic layer approach looks at first sight as if it could be used analytically, so it is necessary to indicate its limitations. The major one is that real species frequently cannot be classified in this way, because their diets are more complex and more specialised than the classification into herbivores and so on suggests. For instance Zwölfer (1971), in discussing the parasitoid insects attacking insect larvae, points out that some species may be both primary parasitoids on the herbivores and also hyper-parasitoids on other parasitoid species attacking the same herbivores. The existence of the system requires that the dual type of parasitoid is less efficient as a primary parasitoid, or it would displace the other species. There are plenty of examples of omnivores that are both carnivores and herbivores. Herbivores may specialize on one part only of a plant, such as the leaves or fruit. Then there are species, such as many soil nematodes, that are specialized bacterial feeders. These are often overlooked in general accounts of communities, and presumably would be classified as carnivores or decomposers. None of these examples fits the simple conventional classification. Another limitation of the trophic layer approach, perhaps less important but certainly troublesome in trying to join population and community analyses, is that species change their trophic habits with age. For instance frogs are herbivores when tadpoles but carnivores when adults, and there are innumerable other examples.

Even with these limitations, the trophic layer concept explains much of the variation in abundance, producing the well known pyramid of numbers. This can be seen in Fig. 16.1 to some extent. Starting at the right of the Figure, line 15 refers to the only primary producers counted, lines 22, 4, 23, 19, 11 and 6 are the dominant, least specialized herbivores, line 2 is a specialized herbivore, and so on traversing left across the Figure on to omnivores and carnivores. However some of the lines towards the left, such as line 9, refer to generalized rare herbivores which are dominant members of the plankton either at other seasons or in other places. Food and productivity may determine the total numbers of, say, herbivores, but there is no reason to suppose that they determine the relative number of species with the same food habits. There appears to be no theory yet developed to explain these differences in mean abundance which has not some severe mathematical drawback. Some of these theories are discussed in MacArthur and Wilson (1967), Pielou (1969), and Cohen (1968). A satisfactory theory will presumably have to be developed from those of two species interactions studied in Section III, bearing in mind the extra complications already discussed that occur when more than two species are involved. So the rest of this Chapter is concerned with the efficient empirical description of the variation in the numbers of different species in a community, and the interpretation of this description. Some other properties of the community will be considered, partly in relation to this description, in the next Chapter.

The data matrix

The basic data on any community can be set out as a table of the numerical abundance of each species against the places or times at which the samples were taken. In Chapter 1, the data considered were the number of a species at one place in each of a set of successive years. In matrix terms, this uses a vector, of one row for one species, and with each column refering to a separate year. The natural extension of this to a community is an $m \times n$ matrix, with m rows for the species and n for the years. As was noted in Chapter 14, when discussing the analysis of seasonal changes in the plankton, it is legitimate and satisfactory to group the species, even though this leads to a loss of some information. Then the rows of the matrix may refer to single species, some stages only of a species, or groups of species. These variables were called, for convenience, entities by Williamson (1961a) and Fig. 16.1 is a plot of the data matrix in that paper. Because there were twenty-three entities counted in the eleven years, the y axis of Fig. 16.1 refers to the rank of each entity independently, and not to the years. In Fig. 16.2, four entities have been plotted against years, to help to give a clearer idea of the content of the data matrix.

In plant ecology in particular, community analyses have been done using presence and absence only in the data matrix. This is a useful preliminary in deciding what samples can be combined together as members of one community, but it is only a preliminary in studying communities as an aid in population analysis, and so I shall not discuss it here.

The data matrix can be regarded as a table in a two way analysis of variance. The variance between rows is the variance between the means of each species, and is a reflection therefore of the mean community structure, its trophic structure and pyramid of numbers, which was considered at the beginning of this Chapter. The variance between columns is the variance between years. It might appear to be in the variance in productivity, but as the species are measured logarithmically, this is not so, a point which will be mentioned again in Chapter 17. The third part of the variance that would normally be considered in an analysis of variance is that within rows and columns. For a community study, the variance between rows has as clear a meaning as can be hoped for from a

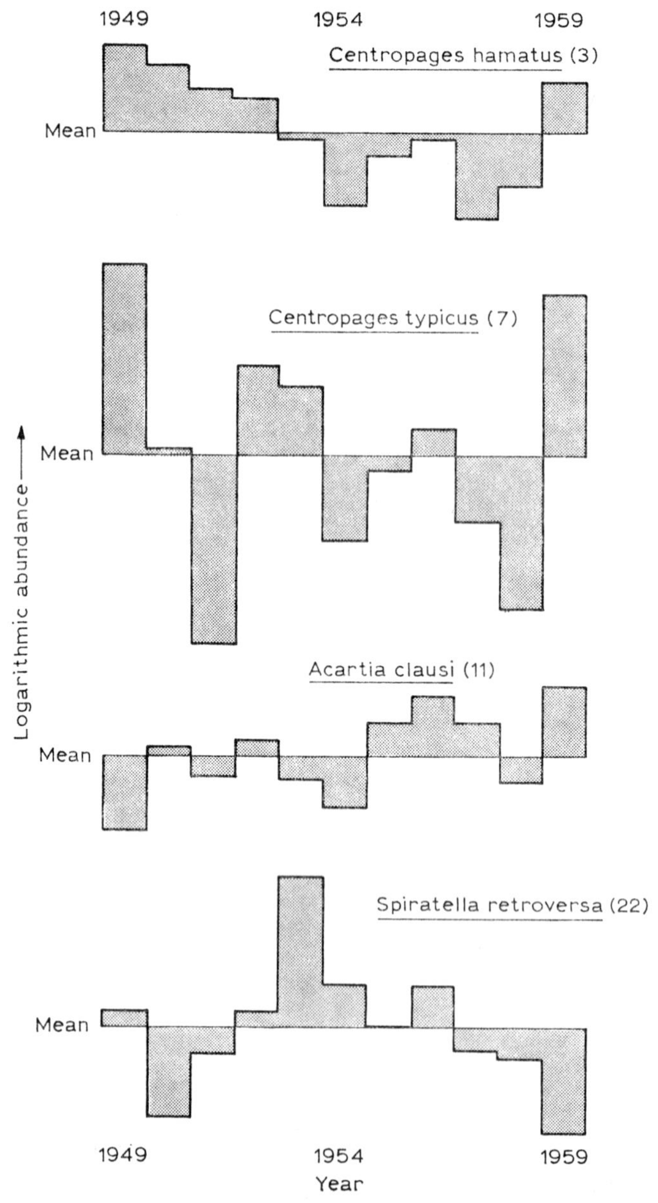

Fig. 16.2 A plot of part of the data matrix of Williamson (1961a). The figure shows the mean abundance in each summer of four entities plotted logarithmically against the year of sampling. *Centropages hamatus* and *C. typicus,* entities 3 and 7, are omnivorous copepods (Gauld 1962), *Acartia clausi* and *Spiratella retroversa* entities 11 and 22 are europhagous herbivores. Acartia is a copepod, Spiratella a pteropod. For comparison with Fig. 16.4, the data are shown as deviations from the mean of each species. These data are also shown in rank order in Fig. 16.1.

statistical study. The variance in the data matrix that needs further analysis is the sum of the other two, and this is the variance within rows, that is the variance of all the species around their own means considered as a set.

Principal Component Analysis

It is well known that in the analysis of variance, a distinct orthogonal component can be associated with each degree of freedom. The object of Principal Component Analysis is to partition the variance within rows of a data matrix into new variables, the Principal Components. There are as many of these as there are rows, but the first Principal Component has the largest variance that can be found by a linear transformation of the data, the second the largest variance orthogonal to the first and so on. It is usual in introducing this topic to refer to the data matrix as defining a cloud of points in a hyper-space. This is easiest to understand by starting with only two dimensions, and Fig. 16.3 shows a cloud

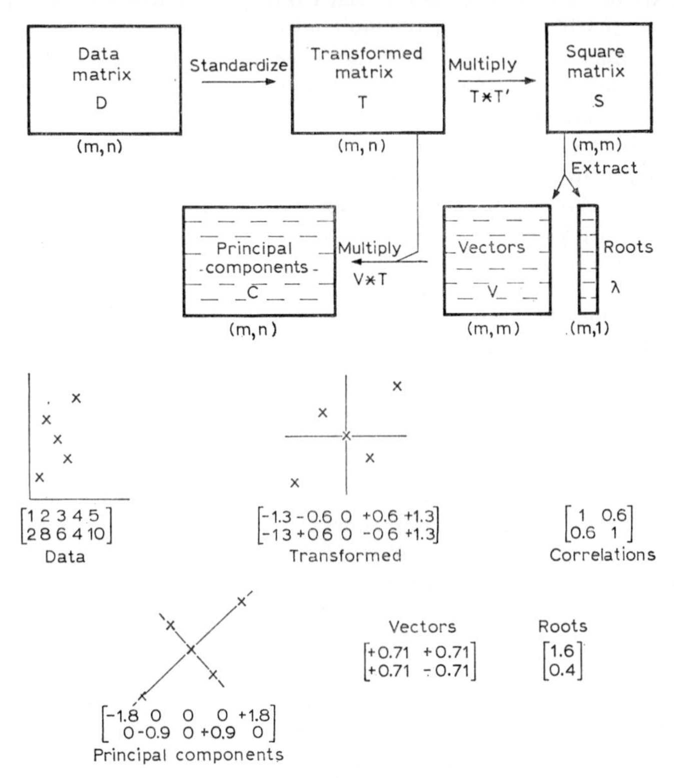

Fig. 16.3 A diagram, with a numerical example, of the processes involved in deriving Principal Components from a data matrix. For a further explanation see the text.

of only five points in a space of two dimensions. These points are defined by the 2×5 matrix also given in the Figure.

There are four steps in calculating Principal Components, and these are shown in Fig. 16.3. They are, first transforming the data to remove the variance that will not be analysed, second forming a symmetrical matrix comparing each variable with every other one, third finding the latent roots and vectors of the symmetrical matrix, and fourth calculating the Principal Components. These will now be considered in more detail.

The first step is to remove the between row variance. This is done by scaling each row to have a zero mean. If the measurements of the rows are not comparable it may also be necessary to standardize the variance of each row. The result of either of these can be called a transformed data matrix. Standardising only the means leads to a covariance matrix at the second step, standardising the variances as well to a correlation matrix. Correlation matrices have been used much more than covariance matrices in Principal Component Analysis. In many applications the rows are not comparable, and indeed I thought that this was so in plankton (Williamson 1961b) because of the ambiguity of using numbers or biomass or even some other measure or the abundance of each species. The argument Lewontin (1966) uses in another context applies here too: if a logarithmic scale is used, all the variances are comparable. So for a Principal Component Analysis of the logarithm of population numbers, a covariance matrix seems preferable. A correlation matrix loses the information on the variance of each species, and this information should be retained unless it is thought, say, to be an artifact of sampling. Multiplying the transformed data matrix by its transpose will give a symmetrical matrix of sums of squares and sums of products, and scaling this by the number of degrees of freedom will give the covariance or the correlation matrix, depending on the standardisation that was used.

In Fig. 16.3 a correlation matrix is shown, as the derivation of this is slightly more complicated. It can be seen that scaling the points has the effect of moving the axes into the centre of the cloud, and, in this case, distorting the cloud so that it has the same variance along both axes. The third and fourth steps rotate these axes, so that the first of them lies along the direction of the greatest variance in the cloud. At the third step, latent roots and vectors are calculated (see the Appendix). The vectors should be standardized so that the sum of the squares of their elements equals 1. The transformed data matrix and the vector matrix are then multiplied together at the fourth step to give the principal components.

The vectors are orthogonal to each other, the sum of the squares of the inner product of any vector with any other is zero. They give the direction of the new axes, and with a 2×2 correlation matrix, these directions are always the two lines with a slope of $+1$ and of -1. The components give the position of the points by reference to these new axes. The variance of each component is given by the corresponding latent root. Although the vectors are orthogonal, they are not necessarily uncorrelated, because they can have any mean, and this will not usually be zero. The components combine the orthogonality of the vectors and the zero mean of the rows of the transformed data matrix, and therefore are uncorrelated.

There are two points about Principal Components which should be mentioned for completeness, but may be skipped by the reader who only wants to know how to apply and interpret them. The first is that if the multiplication of the data matrix by its transpose is done the other way round, the symmetrical matrix will not be a covariance or a correlation matrix, because the columns have not been standardized. However the second symmetrical matrix will have the same roots as the first (any extra roots either way being zero) and its vectors will be the components of the first. If these are standardized so that the sums of the squares equal one, multiplying by the transformed data matrix will give the original vectors, but now the sums of squares of them will equal the latent roots. Multiplying by the transformed data matrix acts as a λ amplifier. This point seems to have been first made by Slater (1958), and though his proof needs a slight extension to cover λ amplifier, this is easily provided (Williamson, in press).

The second point is related. If the data matrix is standardized by columns, then in the ecological context being considered, a covariance matrix of years rather than species would be produced. From the covariance matrix of species the analysis gives vectors in species and components in years, while the covariance of years gives vectors in years and components in species. The cross correspondence will not be perfect, because the initial standardisations are different, but in practice very similar scores result (Williamson 1961b).

The application of Principal Component Analysis

Figure 16.4 shows the first six Principal Components of the correlation matrix of the data in Fig. 16.1. It can be seen that each has a zero mean, and that the variances decrease progressively; it is not perhaps obvious that they are uncorrelated, but this is so. The first use of the component is to show the relationships between the years, and one of many possible plots is shown in Fig. 16.5. This may be considered along with Fig. 16.6, showing two components of species plotted against each other, derived from the covariance matrix of years. Species that are close to each other in Fig. 16.6 have shown similar variations over the eleven years. For instance numbers 1 to 4 have all had population fluctuations like *Centropages hamatus*, which is number 3, in Fig. 16.2, while numbers 18 to 22 inclusive were rather like *Spiratella retroversa*, number 22, also shown in Fig. 16.2. The variation of *C. hamatus* can be compared with the first component in Fig. 16.4, and they are clearly similar. However, the sign on the scale of the first component is arbitrary, as is the sign of the vector it is derived from, so that if, contrary to convention, positive values are read downwards, the first component is also rather like the population curve of *Spiratella retroversa*. The first component summarizes the difference in the population curves of the two species and the others that vary like them. The result of this is seen in Fig. 16.5, where the years fall into two groups. At one end are 1949 to 1952 inclusive and 1959, at the

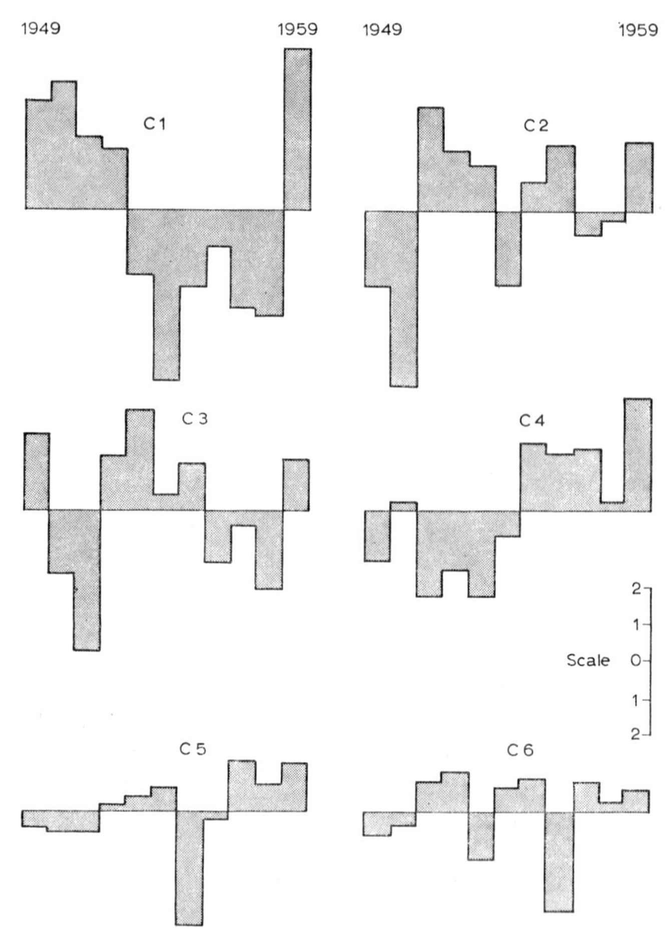

Fig. 16.4 The first six Principal Components of the correlation matrix of entities (species etc.) from the data matrix of Williamson (1961a). The data are shown in Fig. 16.1 and 16.2. The components are uncorrelated, each has a zero mean and the variance is less in each successive component from C1 to C6. The scale is in standard deviations for one entity (from Williamson 1963).

other 1953 to 1958 inclusive. In the first group of years *Centropages hamatus* was on the whole above its average, and *Spiratella retroversa* below its average, and the reverse is true in the second group of years.

The major result of Principal Component Analysis is a classification both of the years and species into sets that behave similarly, and these comparisons may be made along a number of different axes of decreasing importance. This is a useful step in analysis, but it must be emphasised that the components are strictly

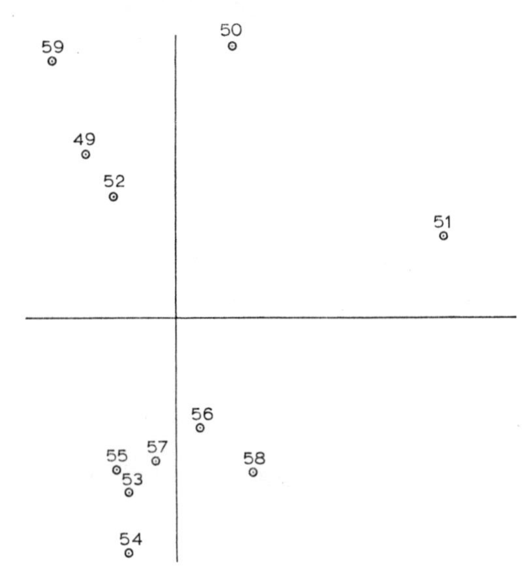

Fig. 16.5 The relation of the years 1949–1959 as measured by the plankton in the north-west North Sea (Fig. 16.1). The axes are scaled by vectors of the covariance matrix of years. The vertical axis seems to be related to the stability of the thermocline; weak thermoclines at the top of the Figure, strong thermoclines at the bottom. The horizontal axis seems to be related to water temperature in spring; warm years to the left, cool years to the right. See also Fig. 16.6.

empirical, they are derived entirely from the data. It is also worth remembering that although the components group together species that have had similar variation in their population sizes, they do not necessarily select species that have particularly close ecological relationships. There is no reason to suppose that in 1954, say, the species at one end of the first component did not interact as strongly with the species at the other end as with each other. All the species are part of the same community. What the components do pick out are the major changes in variability in the community as a set, and most of these are likely either to have been caused by, or at least to have arisen indirectly from, environmental changes.

The interpretation of Principal Components

If Principal Components arise from environmental variation, or for that matter from variation in any other measurable factor, can this relationship be identified? As was noted in Chapter 1, there is a reluctance amongst ecologists to try to identify these relationships for fear of statistical booby traps. There are two reasons why these fears are unnecessary. The first is that what can be derived from such studies are hypotheses and not proven relationships, and the second is that there are supplementary tests that can be used to test these hypotheses. The procedure

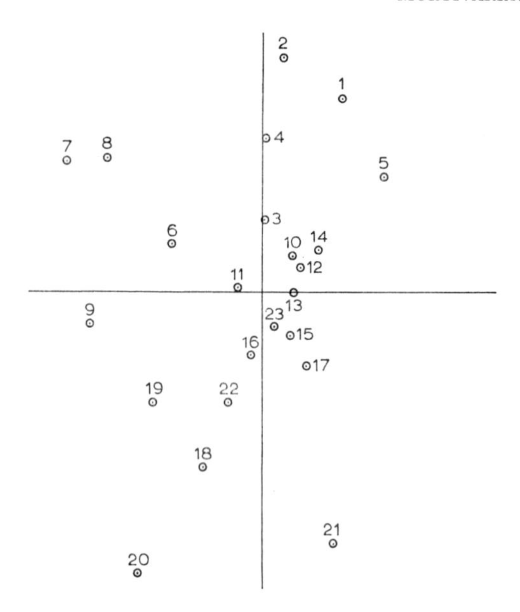

Fig. 16.6 The relation of plankton entities in samples from the north-west North Sea as measured by their variation in abundance in the period 1949–1959. The axes are scaled by components derived from the covariance matrix of years, and correspond to the axes in Fig. 16.5. Entities at the top of the Figure appear to be favoured in years with weak thermoclines, those at the bottom in years with strong thermoclines. Entities to the left seem to be associated with warm springs, those to the right with cool ones. As this Figure is derived from a covariance matrix, the entities that have been more variable tend to occur further out. Entity 20 (*Clione limacina*) has the greatest variance, entity 23 (*Calanus finmarchicus,* stages I-IV) the least (c.f. Table 1.1)

is simply to test the correlation between a principal component and any environmental variable that might be relevant. If this correlation is not significant, then that variable will be set aside. Problems come if the correlation is significant.

Assume for the moment that the significance of the correlation is accepted, what is the position? A hypothesis has been supported by the data and tested on the same data. It is therefore still only a hypothesis, but that does not prevent it from being useful. Research does not stop when a Principal Component Analysis has been completed, and the hypothesis can be used to plan new experiments or observations, on which it then may be legitimately tested. Meanwhile it is a hypothesis that can be discussed. A procedure should not be rejected because it cannot do all that one might like.

One reason sometimes put forward for not using this procedure is that if the component is tested against 20 different environmental variables, one will be significant at the 5% level by chance. In practice this is no problem. It is a lucky ecologist, at least in marine research, who has 20 sensible variables to

F

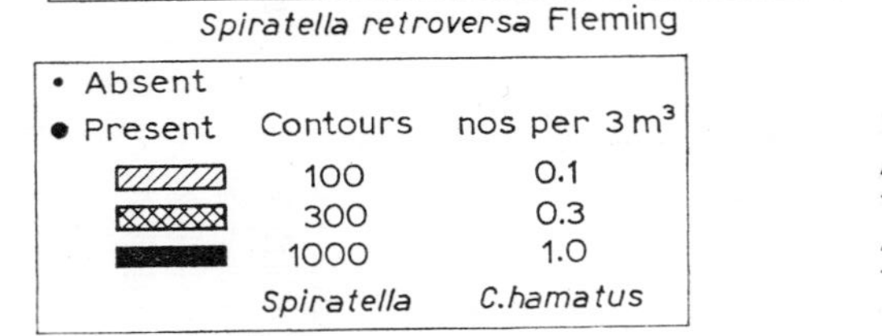

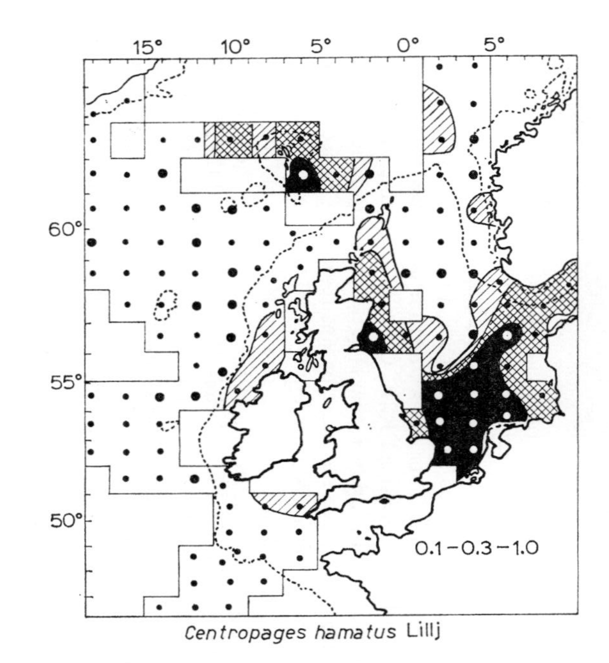

Fig. 16.7 The distribution of *Spiratella retroversa* (entity 22) and *Centropages hamatus* (entity 3) around the British Isles. *S. retroversa* is commonest towards the edge of the continental shelf, and is towards the bottom of Fig. 16.6. *C. hamatus* is commonest in the shallow water of the southern North Sea, and is towards the top of Fig. 16.6. (after Vane 1961 and after Colebrook, John and Brown 1961).

test, and the significance of the correlations is frequently at the 1% level, or even the 0.1% level. However, the other ways of examining a hypothesis generated by a correlation are more important than its formal significance level. Consider the hydrographic function which was found to have a correlation of 0.9 with the first component in the figures. This was a function of the surface temperature anomalies, suggested by the physicist R. E. Craig as a possible measure of the strength of the thermocline in summer, that is of the extent to which the sea was tending to a definite two layered structure or was remaining mixed throughout as in the winter. This function has been found useful too in a Principal Component Analysis of phytoplankton seasonal cycles by Robinson (1970).

Do the groups of species at either end of the component fit the interpretation that the variation is associated with variation in the strength of the thermocline? This is a biological question, and is more powerful than it might seem at first sight, as it is a single tailed test. Not only must the differences between the species be of the right sort, but they must be the right way round. Species associated with weak thermoclines have to do better under this test in years where Craig's anomalies indicate a weak thermocline. It is known that thermoclines are weaker in shallower water, and indeed scarcely discernible in the southern North Sea. So the species groups should be those that are particularly abundant in the southern North Sea compared with those that are more particularly abundant further out towards the edge of the continental shelf. Figure 16.7 shows the distribution of *Spiratella retroversa* and of *Centropages hamatus*, and it can be seen that these are just what they should be if the hypothesis suggested by the correlation is correct.

This sort of test can work unexpectedly well. The other component shown in Figs. 16.5 and 16.6 was thought to be a function of spring temperature (Williamson 1961b) and the species involved, group B of Williamson 1961a, thought not to form a geographical group. In a Principal Component Analysis of distribution, Colebrook (1964) showed that Group B were members of what he calls the southern intermediate group, which had not been recognized earlier, and so fitting the interpretation by temperature.

Principal Components are a powerful way of unravelling the importance of different environmental variables, but the method has important limitations. It is entirely empirical, and uses simple criteria for finding new axes—the largest uncorrelated variances possible. Other criteria, involving other rotations, are used in factor analysis, but there seems to be no advantage in using these in analysing communities, particularly when the end result is only a first hypothesis. The other important limitation is that it is only the variation in population numbers that is considered. This variation is assumed to be related to other measurable variables, and so, from the theory of Chapter 3, it is assumed that the populations are more or less in equilibrium with these other variables. Principal Component Analysis can only suggest hypotheses that explain the variation in these equilibria, not their existence, nor the organization of the community as a whole.

17 Properties of some species systems

Principal Component Analysis of communities shows that mathematical studies of groups of species occurring together can give intelligible and useful results. These results are limited to clarifying the relations between variation in the sizes of the populations and in variation of the environment. For an understanding of the whole biology of a community something more is needed. Much of this is simple biology of the individual species, as discussed in the first two sections of this book, or of the interactions between them, which were discussed in Section III. But there may be properties which depend on large sets of species. Many of these have been suggested in the past, and a partial classified list of them is given in Table 17.1. These are placed, rather arbitrarily, into four groups, with

Table 17.1 Some properties of communities that might be measured

A	B	C	D
Richness	Stratification	Metabolism	Vulnerability
Diversity	Periodism	Energetics	Stability
Complexity	Succession		
	Climax	Success	
	Development	Distribution	
	Latitudinal variation		

'latitudinal variation' linking groups B and C. It may well be that there are important relations between the qualities indicated by these terms, but if so there is one important point which seems to have been neglected: these relations should be demonstrable by ordinary statistical techniques. If, for instance, complexity (in group A) is thought to be related to stability (in group D), then both complexity and stability need be measured, and their relationship studied, by correlation coefficients or in any other accepted way.

How have these groups been derived? Group A are properties related to the number of species, and to the feeding and other interrelations of the species. The number of species is in fact a possible measure of all the terms here, and any other measure should be shown to be better than that. Group B are the classic properties of communities, and these all involve variation either in space or time. Group C are related to the flow of energy through a community, assuming that more widespread or more productive communities are more successful. Group D refer to properties measured on the observed changes of numbers in time. I am indebted to Mr. B. Stableford for pointing out that Group D can only be measured over a period of time, but that Group A can be measured instantaneously, and so are necessarily different. To assume, as MacArthur (1955) did in an interesting pioneering paper, that stability is complexity is to overlook this point.

In groups A, B and D, the numbers in the individual species need to be counted and so it is reasonable to expect that the measure of these properties will be derived from the logarithms of the numbers. As has been argued repeatedly in this book, the logarithm and the rate of change of the logarithm of the numbers of a species are the prime variables in population ecology. Group C is rather different, depending on the total throughput of energy, and so involving, to some extent at least the biomass and numbers measured arithmetically. The measure of community structure which seems most related to population analysis is stability, and much of this Chapter will be devoted to discussing how to measure it. It is widely believed to be related to diversity and complexity, and it will be necessary to mention some work on these. However, taking for the moment the usual view that everyone has an intuitive understanding of these terms, it is worth noting the six reasons that Elton (1958) put forward for thinking that group D is related to Group A. These were:

1 Mathematics of predator/prey systems
2 Laboratory experiments on predator/prey systems } show that simple systems are unstable.
3 Islands, particularly isolated oceanic islands, are readily colonisable.
4 Invasion and outbreaks are much commoner in cultivated land than in natural communities.
5 There are no outbreaks in tropical forests.
6 Pest problems can be severe in heavily sprayed orchards.

As Elton (1966) noted, point 3 concerns vulnerability rather than stability. The discussion in Chapter 14 suggests that points 1 and 2 are far from necessarily true. The other three points seem perhaps to relate to disturbance as much as to complexity, but that is another undefined term. So, provided measurements confirm the stability of the tropical forests, the last three points at the very least imply that there is a hypothesis worth investigating.

The measurement of diversity and complexity

MacArthur's (1955) measure of complexity has been mentioned. This is based on the number of food links in a community. This interesting idea might be rather hard to define in practice, because the food preferences of many species change to the food available. Like the most popular measure of diversity, it is based on the 'information' content of the system. The use of information theory in measuring diversity depends on calculating the probability that the next specimen in a sample will belong to a species new to that sample. The procedure is described for instance by Southwood (1966). Unfortunately as Fager (1963) points out, this information index is sensitive to changes in the proportion of only the few commonest species, and so is unsatisfactory for measuring changes in the community as a whole. This comes from not using logarithmic counts. One of the main conclusions drawn from using the information index of diversity is that during the succession of phytoplankton from spring to summer the community becomes more diverse. As Fogg (1965) says this is a well known phenomenon in plant succession. So for the moment the number of species still seems to be the simplest, the least ambiguous and the most satisfactory measure of complexity and diversity.

The measurement of stability

In Chapter 1 it was argued that the standard deviation of the logarithm of the population size was the best measure of its variability. In Chapter 16 the variability of a community was studied by using the covariance matrix, which has the variances of the logarithm of each population size on its principle diagonal. For a single species the variance and standard deviation are exactly equivalent. For more than one species it is more convenient to work with variances when, as in a covariance matrix, they all have the same number of degrees of freedom, because then they are additive. A more variable population is empirically less stable, though there could be other definitions of the stability of a single population. Can the covariance matrix be used to derive a measure of the stability of a community?

An $n \times n$ covariance matrix has n species variances v_i and n latent roots, λ_i, which are the variances of the principal components. One possible statistic is the sum or mean of the variances, Σv_i or $\Sigma v_i/n$, and this is equal to the same function of the roots $\Sigma\lambda_i$ or $\Sigma\lambda/n$, and the sums are the trace of the matrix (see Appendix). Other possible functions of the latent roots are unsatisfactory. These could be, for instance the determinant of the matrix, which $\Pi\lambda_i$ or $\exp(\Sigma\, ln\, \lambda_i)$ or the variance of the roots or of the logarithms of the roots, var (λ_i) or var $(ln\, \lambda_i)$. All these measures are sensitive to the smallest roots, and these roots are, in any real matrix, almost entirely determined by sampling errors and even rounding errors in the data. The only measures other than the trace of the matrix that could be based on the roots seem to be either the size of the dominant root λ_1, or possibly the ratio of this to the trace, as these measure the amount or proportion of the variance ascribable to the most important environmental variable, or the way in which the community is most sensitive to the environment.

There is one other measure that might be suggested, and that is the sum of all the elements of the covariance matrix. Remembering that

$$\text{var}\,(A+B)=\text{var}\,(A)+\text{var}\,(B)+2.\ \text{covar}\,(A, B),$$

the sum of the matrix is clearly the variance of the combined measure of all the species, though with each species measured as its deviation from its own mean.

Even three possible measures, the trace, λ_1 and $\Sigma\Sigma c_{ij}$, are enough to show that for a community there may be more than one type of stability. Some trial calculations I have done show that the correlation between these three may be small, and that none of them need be strongly correlated with the number of species in the community. This work will be published later.

One or more of these measures may well prove to be a satisfactory measure of some important aspect of community stability. My reason for saying this is that Watt (1964, 1965) has found some interesting results using the variability of single populations, using in fact the standard error of the logarithm of the population size. He found

1 that the stability of the members of a trophic layer increases as the number of species in that layer increases,
2 that the stability of a species decreases as the number of species attacking it increases,

and 3 that the stability of a species decreases as the proportion of the environment that is its food increases.

His second conclusion, in particular, is not what might be expected on a simple relation of complexity to stability, and, as was noted in Chapter 14, has an important bearing on the practice of biological control.

The analysis of communities is particularly challenging, because it involves the analysis at the same time of single populations and of the interactions between different species, and because it is in communities that evolution takes place. The techniques of analysis that have been developed should lead to a better understanding of the properties of communities before very long.

Appendix. Summary of matrix manipulations

This is a brief survey only. Those who have not met matrices before should consult a work such as Searle (1966), Smith (1969) or Bickley and Thompson (1964). Some understanding of complex numbers and the theory of equations is also helpful, and short notes on these are included at the end, with a definition of the modulus of a number.
A matrix is a rectangular array of numbers, arranged in rows and columns:

$$\begin{bmatrix} 5 & 7 & 3 \\ 2 & -9 & 0 \end{bmatrix}$$ is a two row by three column, or a 2×3, or a (2,3) matrix.

Each number is termed an element. Usually a matrix is designated by a capital bold-face letter, such as $\mathbf{A}$, and the elements by lower case letters with double subscripts such as a_{ij}, where $_i$ refers to the row and $_j$ to the column in which the element is found.

Special matrices

A matrix with the same number of rows and columns, $n \times n$, is a square matrix. A matrix with only one row, or one column, $(1,n)$ or $(n,1)$ is a row or column vector, often designated by a lower case bold letter such as $\mathbf{y}$, and the elements need only one subscript. A square matrix with ones on the principal diagonal and zero elements elsewhere is the unit matrix, designated

$\mathbf{I}$ e.g. $\begin{bmatrix} 1 & 0 \\ 0 & 1 \end{bmatrix}$ is the 2×2 unit matrix.

The trace of a matrix is the sum of the elements of the principal diagonal:
$\mathrm{Tr}\,(\mathbf{A}) = \Sigma\, a_{ii}$, and the trace of a unit matrix of order n is n.
A null matrix has zero for all the elements.
A diagonal matrix is one with non-zero elements only on the principal diagonal.
A symmetrical matrix is one in which the elements below the principal diagonal are a reflection, through that diagonal, of the elements above. It is therefore unchanged by transposition (see below).

Matrix operations

Addition and snbtraction. If the number of rows and columns are the same in two matrices, they are added (or subtracted) by adding (or subtracting) the corresponding elements:

$$[5 \quad 2] + [-1 \quad 3] = [4 \quad 5]$$

Transposition. Each row becomes a column (and so each column becomes a row), so the matrix is flipped about its principal diagonal. The operation is usually written $\mathbf{A}^T$, sometimes $\mathbf{A}'$. Under transposition, a column vector becomes a row vector, and vice versa. If a square matrix is unchanged under transposition, i.e. the elments of the ith row are the same as the elements of the ith column, the matrix is symmetrical.

Scalar multiplication. A matrix multiplied by a single number, a scalar, has all its elements multiplied by the scalar

$$k \times [a \quad b] = [ka \quad kb].$$

Multiplication. Matrix multiplication is by rows and columns, with summation. The number of columns in the first matrix must be equal to the number of rows in the second. So the dimensions of the operation are

$$(m,n) \times (n,p) \rightarrow (m,p)$$

$$\begin{bmatrix} a & b \\ c & d \end{bmatrix} \times \begin{bmatrix} e & f \\ g & h \end{bmatrix} = \begin{bmatrix} ae+bg & af+bh \\ ce+dg & cf+dh \end{bmatrix}$$

$$2 \times 2 \qquad\qquad 2 \times 2 \qquad\qquad 2 \times 2$$

A column vector pre-multiplied by a square matrix produces a column vector

$$(n,n) \times (n,1) \rightarrow (n,1).$$

Inversion. The matrix equivalent of division. With numbers, if $xy=1$, then $y=1/x=x^{-1}$. With matrices, when $\mathbf{A}$ and $\mathbf{B}$ are both square matrices, if $\mathbf{AB}=\mathbf{I}$ then $\mathbf{B}=\mathbf{A}^{-1}$, $\mathbf{A}=\mathbf{B}^{-1}$ and $\mathbf{BA}=\mathbf{I}$ also.

The inverse of any matrix more than 2×2 is usually found by a computer program. If the determinant (see below) of a matrix is zero, or, what is the same, if any one of the latent roots (see below) is zero, the matrix has no inverse.

$(\mathbf{A} \times \mathbf{B} \times \mathbf{C})^{-1} = \mathbf{C}^{-1} \times \mathbf{B}^{-1} \times \mathbf{A}^{-1}$ and also $(\mathbf{A} \times \mathbf{B} \times \mathbf{C})^T = \mathbf{C}^T \times \mathrm{B}^T \times \mathbf{A}^T$.

Determinant. This is a function of the elements of a square matrix. If the elements are numbers, then the determinant is a single number. It is designated $|\mathbf{A}|$ or det $[\mathbf{A}]$. For a 2×2 matrix $\begin{bmatrix} a & b \\ c & d \end{bmatrix}$ the determinant is $(ad-bc)$, two terms each of two elements. Remembering that $2!=2$, for an $n \times n$ matrix, the determinant is a function of $n!$ terms each of n elements, and in any one term each element comes from a different row and column from all the other elements of that term. As $n!$ increases rapidly with n, determinants are not usually calculated for any but small matrices.

Latent roots and vectors (Eigen values and eigen vectors). If, for a square matrix, a vector can be found such that multiplication by the matrix is equivalent to multiplication by a scalar, then that vector is a latent vector (or eigen vector) and the scalar is a latent root (or eigen value), i.e. if

$$\mathbf{Ax} = \lambda\mathbf{x}$$

where $\mathbf{A}$ is $n \times n$, $\mathbf{x}$ is a column vector, $n \times 1$, and λ a number, then λ is a latent root of $\mathbf{A}$ and $\mathbf{x}$ the corresponding latent vector. In this case, it will also be true that $c\,\mathbf{Ax} = \mathbf{A}(c\mathbf{x}) = c\,\lambda\mathbf{x} = \lambda\,(c\mathbf{x})$, where c is any scalar, so the elements of $\mathbf{x}$ are only defined relative to each other, not in absolute scale.

For instance the matrix $\begin{bmatrix} 17 & 3 \\ 3 & 9 \end{bmatrix}$ has the latent roots 18 and 8, and the latent vectors $\begin{bmatrix} 3 \\ 1 \end{bmatrix}$ and $\begin{bmatrix} -1 \\ 3 \end{bmatrix}$

because $\begin{bmatrix} 17 & 3 \\ 3 & 9 \end{bmatrix} \begin{bmatrix} 3 \\ 1 \end{bmatrix} = \begin{bmatrix} 54 \\ 18 \end{bmatrix}$ and $\begin{bmatrix} 17 & 3 \\ 3 & 9 \end{bmatrix} \begin{bmatrix} -1 \\ +3 \end{bmatrix} \quad \begin{bmatrix} -8 \\ +24 \end{bmatrix}$

The elements of $\mathbf{x}$ may be standardized in a number of conventional ways, particularly

$$\text{either } \Sigma x_i^2 = 1 \text{ or } \max\,[x_i] = 1.$$

An $n \times n$ matrix has n latent roots, λ_i which are also the roots of the 'Characteristic equation' of the matrix which is the determinantal equation

$$|\mathbf{A} - \lambda\mathbf{I}| = 0,$$

which can be expanded to the nth order polynomial

$$\lambda^n + c_1\lambda^{n-1} + \ldots c_{n-1}\,\lambda + c_1 = 0$$

where the c_i are the coefficients derived from the elements of the matrix. This equation has n roots, the n latent roots of $\mathbf{A}$, and these may be real (positive or negative), complex or zero. Each non-zero root has a distinct latent vector associated with it, and elements of the vector may be real, complex or zero. (In theory, with non-symmetrical matrices it is possible for two or more vectors to be identical, but such cases do not occur in practice in population studies).

The sum of the roots = the trace of the matrix

$$Tr\,(\mathbf{A}) = \Sigma a_{ii} = \Sigma\lambda = (-1)^1 c_1$$

The product of the roots=the determinant of the matrix

$$[\mathbf{A}]=\Pi\lambda=(-1)^n c_n$$

A symmetrical matrix has only real (positive, zero or negative) roots, and the vectors consist of real elements.
The root of greatest modulus is termed the dominant latent root, and its vector the dominant latent vector.
If λ and $\mathbf{v}$ are a root and vector of $\mathbf{A}$, the λ^n is a root of $\mathbf{A}^n$ and $\mathbf{v}$ (not $\mathbf{v}^n$) is the corresponding vector when n is any real number, positive or negative.

Complex and imaginary numbers, modulus

Real numbers may be represented as points on a line, conventionally a horizontal line with positive numbers to the right and negative to the left. Complex numbers are ordered couples of real numbers, such as (a,b) where a and b are real, and are often referred to by a single symbol, such as $Z\equiv(a,b)$. By a natural extension of representing real numbers on a line, complex numbers may be represented as points in a plane (see the Fig.) Elements on the real (first) axis may be transferred to the imaginary (second) axis by multiplication by the operator i, e.g. $(a,0)\times i=(0,a)$.
So a natural notation for the complex number $(a,b)=Z$ is $(a+bi)$. $a+bi$ is a complex number, $a+0i$ is a real one, and $0+bi$ an imaginary number.

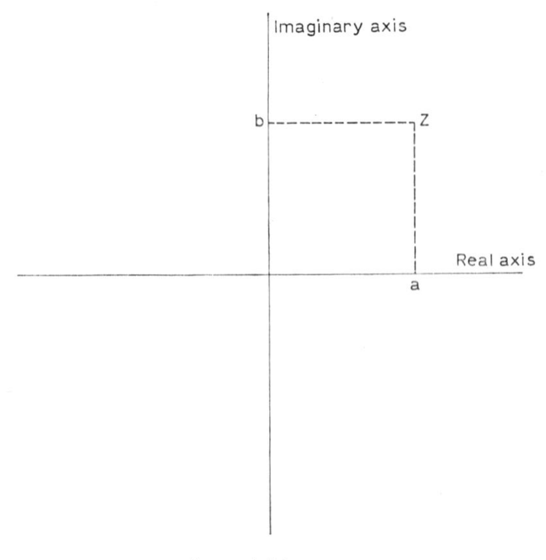

Argand Diagram.

The operator i transfers a number 90° anti-clockwise, $i \times i$ converts a number to its negative,

$$a \times i \times i = ai^2 = -a = -1.a,$$

so i is the square root of minus one. The two dimensional diagram for complex numbers is called an Argand diagram.

The modulus of any number is its distance from the origin in the Argand diagram, so, if $Z = a + bi$ is any complex number (real if $b = 0$, a and b may be positive or negative) the modulus of $Z \equiv |Z| \equiv \sqrt{(a^2 + b^2)}$, by Pythagoras. Real, imaginary and complex numbers may also be represented by 2×2 matrices

$$\begin{bmatrix} 0 & -1 \\ 1 & 0 \end{bmatrix} = \text{J, say, is equivalent to } i \text{, in that } \text{J}^2 = \begin{bmatrix} -1 & 0 \\ 0 & -1 \end{bmatrix} = -\text{I} = -1 \begin{bmatrix} 1 & 0 \\ 0 & 1 \end{bmatrix}$$

So a complex number, $a + bi = Z$, may also be written $\begin{bmatrix} a & -b \\ b & a \end{bmatrix} = Z$, say,

and the determinant of $Z \equiv \det(Z) = (a^2 + b^2) = [Z]^2$, the square of the modulus of Z.

The theory of equations (A few points that may be helpful).

A polynomial equation such as

$$b_o x^n + b_1 x^{n-1} + b_2 x^{n-2} + \ldots\ldots + b_{n-1} x + b_n = 0,$$

where the b_i are constant coefficients, has n roots. That is, there are n numbers, which may be real, complex, or imaginary, that make the equation true. If the roots are r_i, then

$(x - r_1)(x - r_2)(\ldots\ldots\ldots\ldots\ldots\ldots)(x - r_n) = 0$, is another form of the equation. If all the b_i are real numbers, then any complex roots occur in conjugate pairs, as mirror images above and below the real axis: if $(d + ei)$ is a root, so is $(d - ei)$. In biological examples, complex numbers seldom need to be considered but it is necessary to know of their existence, and their occurrence as roots of polynomials in dealing with Leslie matrices, and for finding $q = 0$ from $q = f(q)$, where $f(q)$ is a polynomial function, or the ratio of two polynomial functions, of q (see Chapter 6).

The b_i are functions of the roots r_i:

$$(-1)^1 b_1 / b_o = \Sigma r_i$$
$$(-1)^2 b_2 / b_o = {}_{j>i}\Sigma r_i r_j$$

etc. and hence

$$(-1)^n b_n / b_o = \Pi r_i.$$

References

ALLEE, W. C., EMERSON, A. E., PARK, O., PARK, T. and SCHMIDT, K. P. 1949. *Principles of animal ecology*. Saunders, Philadelphia and London.

ALLISON, A. C. 1964. Polymorphism and natural selection in human populations. *Cold Spring Harb. Symp. quant. Biol.*, **29**, 137–49.

ANDREWARTHA, H. G. and BIRCH, L.C. 1954. *The distribution and abundance of animals*. University of Chicago Press, Chicago.

ANON. 1970. Report of the working group on Atlanto-Scandian herring. I.C.E.S. Cooperative Research Report (A), **17**, 1–43.

ARNOLD, R. 1969. The effects of selection by climate on the land-snail *Cepaea nemoralis* (L.) *Evol.*, **23**, 370–78.

ARNOLD, R. 1970. A comparison of populations of the polymorphic land snail *Cepaea nemoralis* (L.) living in a lowland district of France with those in a similar district in England. *Genetics*, **64**, 589–604.

ARNOLD, R. In preparation.

BAILEY, N. T. J. 1957. *The mathematical theory of epidemics*. Griffin and Co. London.

BAILEY, N.T.J. 1967. *The mathematical approach to biology and medicine*. Wiley, London, New York, Sydney.

BALTENSWEILER, W. 1968. The cyclic population dynamics of the grey larch tortrix, *Zeiraphera griseana* Hübner (=*Semasia diniana* Guenee) (Lepidoptera: Tortricidae). *Symp. R. ent. Soc. Lond.*, **4**, 88–97.

BARKER, J. S. F. 1967. The fitness of single species populations of Drosophila. *Evol.* **21**, 606–19.

BARTLETT, M. S. 1957. On theoretical models for competitive and predatory biological systems. *Biometrika*, **44**, 27–42.

BARTLETT, M. S. 1960. *Stochastic population models in ecology and epidemiology*. Methuen, London.

BEARDMORE, J. 1970. Ecological factors and the variability of gene-pools in *Drosophila. Evolutionary Biology*, supplement, 299–314.

BEAVER, R. A. 1967. The regulation of population density in the bar beetle *Scolytus scolytus* (F.) *J. Anim. Ecol.*, **36**, 435–51.

BEVERTON, R. J. H. and HOLT, S. J. 1957. On the dynamics of exploited fish populations. *Fish. Invest., London*, **19**, 7–533.

BEVERTON, R. J. H. 1962. The long-term dynamics of certain North Sea fish populations. *Symp. Br. Ecol. Soc.*, **2**, 242–59.

BICKLEY, W. G. and THOMPSON, S. R. H. G. 1964. *Matrices: their meaning and manipulation*. Unibooks, London.

BIRCH, L. C. 1962. Stability and instability in natural populations. *New Zealand Science Review*, **20**, 9–14.

BIRCH, L. C. and EHRLICH, P. R. 1967. Evolutionary history and population biology. *Nature, Lond.* **214**, 349–52.

BRISTOWE, W. S. 1935. Further notes on the spiders of the Scilly Islands. *Proc. zool. Soc. Lond.*, 219–32.

BRISTOWE, W. S. 1939. *The comity of spiders*. Ray Society, London.

BROADHEAD, E. and WAPSHERE, A. J. 1966. *Mesopsocus* populations on Larch in England. The distribution and dynamics of two closely related coexisting species of Psocoptera sharing the same food resource. *Ecol. Mongr.*, **36**, 327–88.

BRUCK, D. 1957. Male segregation ratio advantage as a factor in maintaining lethal alleles in wild populations of house mice. *Proc. Nat. Acad. Sci. USA*, **43,** 152–8.

CAIN, A. J., CAMERON, R. A. D. and PARKIN, D. T. 1969. Ecology and variations of some helicid snails in northern Scotland. *Proc. malac. Soc. London.*, **38,** 269–99.

CAIN, A. J. and CURREY, J. D. 1963a. Area effects in Cepaea. *Phil. Trans. R. Soc. Ser. Ser. B*, **246,** 1–81.

CAIN, A. J. and CURREY, J. D. 1963b. The causes of area effects. *Heredity*, **18,** 467–71.

CAIN, A. J. and CURREY, J. D. 1968. Studies on Cepaea. III. Ecogenetics of a population of *Cepaea nemoralis* (L.) subject to strong area effects. *Phil. Trans. R. Soc. B*, **253,** 447–82.

CAIN, A. J. and SHEPPARD, P. M. 1950. Selection in the polymorphic land snail *Cepaea nemoralis. Heredity*, **4,** 275–94.

CAIN, A. J. and SHEPPARD, P. M. 1954. The theory of adaptive polymorphism. *Am. Nat.*, **88,** 321–6.

CAIN, A. J., SHEPPARD, P. M. and KING, J. M. B. 1968. Studies on *Cepaea*. I. The genetics of some morphs and varieties of *Cepaea nemoralis* (L.) *Phil. Trans. R. Soc. Ser. B*, **253,** 383–96.

CANNINGS, C. and EDWARDS, A. W. F. 1968. Natural selection and the de Finnetti diagram. *Ann. hum. Genet. Lond.*, **31,** 421–8.

CANNINGS, C. 1969. A graphical method for the study of complex genetical systems with special reference to equilibria. *Biometrics*, **25,** 747–54.

CARTER, M. A. 1968. Studies on Cepaea II. Area effects and visual selection in *Cepaea nemoralis* (L.) and *Cepaea hortensis. Phil. Trans. R. Soc. Ser. B*, **253,** 397–446.

CLARKE, B. 1962. Balanced polymorphism and the diversity of sympatric species. *Syst. Ass. Publ.*, **4,** 47–70.

CLARKE, B. 1966. The evolution of morph-ratio clines. *Am. Nat.* **100,** 389–402.

CLARKE, B. 1969. The evidence for apostatic selection. *Heredity*, **24,** 347–52.

CLARKE, B. and MURRAY, J. 1969. Ecological genetics and speciation in land snails of the genus *Partula. Biol. J. Linn. Soc.*, **1,** 31–42.

CLATWORTHY, J. N. and HARPER, J. L. 1962. The comparative biology of closely related species living in the same area. V. Inter- and intraspecific inteference within cultures of *Lemna* spp. and *Salvinia natans. J. exp. Bot.*, **13,** 307–24.

COHEN, J. E. 1968. Alternative derivations of a species abundance relation. *Am. Nat.*, **102,** 165–72.

COLE, L. C. 1960. Competitive exclusion, *Science*, **132,** 348–9 and 1675–6.

COLEBROOK, J. M. 1964. Continuous plankton records. A principal component analysis of the geographical distribution of zooplankton. *Bull. mar. Ecol.*, **6** (3), 78–100.

COLEBROOK, J. M., JOHN, D. E. and BROWN, W. W. 1961. Continuous plankton records: contributions towards a plankton atlas of the north-eastern Atlantic and the North Sea. Part II: Copepods, *Bull. mar. Ecol.*, **5** (42), 90–7.

COLEBROOK, J. M. and ROBINSON, G. A. 1965. Continuous plankton records: seasonal cycles of phytoplankton and copepods in the north-eastern Atlantic and the North Sea. *Bull. mar. Ecol.*, **6** (5), 123–39.

COMFORT, A. 1964. *Ageing: the biology of senescence.* Routledge, London.

CONNELL, J. H. 1961. The influence of interspecific competition and other factors on the distribution of the barnacle *Chthalamus stellatus. Ecology*, 710–23.

COOK, L. M. 1959. The distribution in Britain of the scarlet tiger moth, *Callimorpha (Panaxia) dominula* L. *Entomologist*, **92,** 232–6.

COOK, L. M. 1967. The genetics of *Cepaea nemoralis. Heredity*, **22,** 397–410.

CROMBIE, A. C. 1947. Interspecific competition. *J. Anim. Ecol.*, **16,** 44–73.

CURREY, J. D., ARNOLD, R. W. and CARTER, M. A. 1964. Further examples of variation of populations of *Cepaea nemoralis* with habitat. *Evol.*, **18,** 111–7.

CUSHING, D. H. 1964. The work of grazing in the sea. *Symp. Br. Ecol. Soc.*, **4**, 207–25.
D'ANCONA, U. 1954. The struggle for existance. *Bibliotheca Biotheoretia* (*D*), **6**, 1–274.
DARWIN, C. 1859. *The origin of species*. 1st Ed. Murray, London.
DE BACH, P. 1966. The competitive displacement and coexistance principles. *A. Rev. Ent.*, **11**, 183–212.
DE WIT, C. T. 1960. On competition, *Versl. Landbouwk. Onderz.*, **66**, (8) 1–82.
DE WIT, C. T., TOW, P. G. and ENNIK, G. C. 1966. Competition between legumes and grasses. *Versl. landbouwk. Ondenz.*, **687**, 1–30.
DELBRUCK, M. 1949. A physicist looks at biology. *Trans. Conn. Acad. Arts. sci.*, **38**, 173–90.
DEN BOER, P. J. and GRADWELL, G. R. 1971. (Eds.) Dynamics of numbers in populations. *Proceeds of the Advanced Study Institute, Oosterbeek, Netherlands, September 1970*. Pudoc, Wageningen, Netherlands.
DOBZHANSKY, T. 1951. *Genetics and the origin of species*. 3rd. Ed. Columbia University Press, New York and London.
DOBZHANSKY, T. 1957. Mendelian populations as genetic systems. *Cold Spring Harb. Symp. quant. Biol.*, **22**, 385–93.
DOBZHANSKY, T. 1958, Genetics of natural populations. XXVII. The genetic changes in populations of *Drosophila pseudoobscura* in the American southwest. *Evol.* **12**, 385–401.
DOBZHANSKY, T., HUNTER, A. S., PAVLOVSKY, O., SPASSKY, B. and WALLACE, B. 1963. Genetics of natural populations XXXI. Genetics of an isolated marginal population of *Drosophila pseudoobscura*. *Genetics*, **48**, 91–103.
DOBZHANSKY, T., LEWONTIN, R. C. and PAVLOVSKY, O. 1964. The capacity for increase in chromosomally polymorphic and monomorphic populations of *Drosophila pseudoobscura*. *Heredity*, **19**, 597–614.
DUNN, L. C. 1960. Variations in the transmission ratios of alleles through egg and sperm in *Mus musculus*. *Am. Nat.*, **94**, 385–93.
DUNN, L. C. and BENNETT, D. 1967. Maintenance of gene frequency of a male sterile, semi-lethal T-allele in a confined population of wild mice. *Am. Nat.*, **101**, 535–8.
EBERHARDT, L. L. 1970. Correlation, regression, and density dependence. *Ecology*, **51**, 306–10.
EHRLICH, P. R. and RAVEN, P. H. 1964. Butterflies and plants: a study in co-evolution. *Evol.*, **18**, 586–608.
ELTON, C. S. 1958. *The ecology of invasions by animals and plants*. Methuen, London.
ELTON, C. S. 1966. *The pattern of animal communities*. Methuen, London.
ELTON, C. and NICHOLSON, M. 1942. The ten year cycle in numbers of the Lynx in Canada. *J. Anim. Ecol.*, **11**, 215–44.
EPPLEY, R. W. and STRICKLAND, J. D. H. 1966. Kinetics of marine phytoplankton growth. *Adv. Microbiol. Sea*, **1**, 23–62.
EVANS, F. C. and SMITH, F. E. 1952. The intrinsic rate of natural increase for the human louse, *Pediculus humanus* L. *Am. Nat.*, **86**, 299–310.
FAGER, E. W. 1963. Communities of organisms pp. 415–37. in *The Sea* Vol. 2, ed. M. N. Hill, Wiley, New York, London and Sydney.
FELLER, W. 1940. On the logistic law of growth and its empirical verifications in biology. *Acta biotheor. A*, **5**, 51–66.
FELLER, W. 1967. On fitness and the cost of natural selection. *Genet. Res. Camb.*, **9**, 1–15.
FINNEY, D. J. 1952. The equilibrium of a self-incompatible species. *Genetica*, **26**, 33–64.
FISHER, J. and LOCKLEY, R. M. 1954. *Sea birds*, Collins, London.
FISHER, J. and VEVERS, H. G. 1944. The breeding distribution, history and population of the North Atlantic gannet (*Sula bassana*). *J. Anim. Ecol.*, **13**, 49–62.

FISHER, R. A. 1931. *The genetical theory of natural selection.* Oxford U.P.

FOGG, G. E. 1965. *Algal cultures and phytoplankton ecology.* University of Wisconsin Press, Madison and Milwaukee.

FORD, E. B. 1945. Polymorphism. *Biol. Rev.* **20,** 73–88.

FORD, E. B. 1964. *Ecological Genetics.* Methuen and Co. Ltd., London.

FORD, E. B. and SHEPPARD, P. M. 1969. The *medionigra* polymorphism of *Panaxia dominula. Heredity,* **24,** 561–69.

GAULD, D. T. 1964. Feeding in planktonic copepods. *Symp. Br. Ecol. Soc.,* **4,** 239–45.

GAUSE, G. F. 1934. *The struggle for existance.* Reprinted 1964. Hafner Publishing Co. New York.

GAUSE, G. F. 1935. Vérifications expérimentals de la théorie mathematique de la lutte pour la vie. *Actualites Sci. Indus.,* **227.**

GAUSE, G. F. 1937. Experimental populations of microscopic organisms. *Ecol.,* **18,** 173–9.

GAUSE, G. F. 1970. Criticism of invalidation of principle of competitive exclusion. *Nature,* Lond., **227,** 89.

GIBB, J. 1954. Population changes of titmice, 1947–1951. *Bird Study,* **1,** 40–8.

GIBB, J. A. and BETTS, M. M. 1963. Food and food supply of nestling tits (*Paridae*) in Breckland pine. *J. Anim. Ecol.,* **32,** 489–533.

GILBERT, O., REYNOLDSON, T. B. and HOBART, J. 1952. Gause's hypothesis: an examination. *J. Anim. Ecol.,* **21,** 310–12.

GOODHART, C. B. 1962. Variation in a colony of the snail *Cepaea nemoralis* (L.) *J. Anim. Ecol.,* **31,** 207–37.

GOODHART, C. B. 1963. 'Area effects' and non-adaptive variation between populations of *Cepaea* (mollusca). *Heredity,* **18,** 459–65.

GREENWOOD, J. J. D. 1968. Coexistance of avian congeners on islands. *Am. Nat.,* **102,** 591–2.

HAIRSTON, N. G. 1958. Observations on the ecology of Paramecium, with comments on the species problem. *Evol.,* **12,** 440–50.

HALDANE, J. B. S. 1924. A mathematical theory of natural and artificial selection. *Trans. Camb. Phil. Soc.,* **23,** 19–41.

HALDANE, J. B. S. 1948. Biology and Marxism. *Modern Quarterly* (*NS*), **3,** 2.

HALDANE, J. B. S. 1953. Animal populations and their regulation. *New Biology,* **15,** 9–24.

HALDANE, J. B. S. 1954. The statics of evolution. In: *Evolution as a process.* Ed. J. Huxley, A. C. Hardy, E. B. Ford. Allen and Unwin, London.

HAMILTON, W. D. 1964. The genetical evolution of social behaviour. *J. theor. Biol.* **7,** 1–16.

HANCOCK, D. A. 1971. The role of predators and parasites in a fishery for the mollusc *Cardium edule.* In: Den Boer and Gradwell, 1971.

HANCOCK, D. A. and URQUHART, A. E. 1965. The determination of natural mortality and its causes in an exploited population of cockles. *Fishery Invest. Series I. Lond.,* **24,** 1–39.

HARDIN, G. 1960. The competitive exclusion principle. *Science,* **131,** 1291–7.

HARPER, J. L. 1967. A Darwinian approach to plant ecology. *J. Ecol.,* **55,** 247–270, *J. Anim. Ecol.,* **36,** 495–518, *J. appl. Ecol.,* **4,** 267–90.

HARPER, J. L. 1968. The regulation of numbers and mass in plant populations. In Lewontin, (1968a), 139–58.

HARPER, J. L. 1969. The role of predation in vegetational diversity. *Brookhaven Symp. Biol.,* **22,** 48–61.

HARPER, J. L., CLATWORTHY, J. N., MCNAUGHTON, I. H. and SAGAR, G. R. 1961. The evolution and ecology of closely related species living in the same area. *Evol.,* **15,** 209–27.

HASSELL, M. P. 1966. Evaluation of parasite or predator responses. *J. Anim. Ecol.*, **35,** 65–75.

HASSELL, M. P. and VARLEY, G. C. 1969. New inductive population model for insect parasites and its bearing on biological control. *Nature, Lond.*, **223,** 1133–7.

HASSELL, M. P. 1971. Parasite behaviour as a factor contributing to the stability of insect host-parasite interactions. In: Den Boer and Gradwell, 1971.

HERBERT, D., ELSWORTH, R. and TELLING, R. C. 1956. The continuous culture of bacteria; a theoretical and experimental study. *J. gen. Microbiol.*, **14,** 601–22.

HOLLING, C. S. 1968. The tactics of a predator. *Symp. R. ent. Soc. Lond.*, **4,** 47–58.

HOLT, S. J. 1962. The application of comparative population studies to fishing biology—an exploration. *Symp. Br. ecol. Soc.*, **2,** 51–59.

HUFFAKER, C. B. 1971. The phenomenon of predation and its role in nature. In: Den Boer and Gradwell, 1971.

HUFFAKER, C. B., KENNETT, C. E., MATSUMOTO, B. and WHITE, E. G. 1968. Some parameters in the role of enemies in the natural control of insect abundance. *Symp. R. ent. Soc. Lond.*, **4,** 59–75.

HUTCHINSON, G. E. 1951. Copepodology for the ornithologist. *Ecol.*, **32,** 571–7.

HUTCHINSON, G. E. 1953. The concept of pattern in ecology. *Proc. Acad. nat. Sci. Philad.*, **105,** 1–12.

HUTCHINSON, G. E. 1965. *The ecological theatre and the evolutionary play.* Yale University Press.

IVLEV, V. S. 1960. On the utilization of food by planktophage fishes. *Bull. math. Biophys.*, **22,** 371–89.

IWAO, S. and WELLINGTON, W. G. 1970. The western tent caterpillar. Qualitative differences and the action of natural enemies. *Res. Popul. Ecol.*, **12,** 81–99.

JOHNSON, W. E. 1965. On mechanisms of self-regulation of population abundance in *Oncorhyncus nerka. Mitt. int. Ver. Limnol.*, **13,** 66–87.

KEMPTHORNE, O. 1957. *An introduction to genetic statistics.* Wiley, New York and London.

KENYON, K. W., SCHEFFER, V. B., and CHAPMAN, D. G. 1954. A population study of the Alaska fur-seal herd. U.S. Fish and Wildlife Service, Special Scientific Report—*Wildlife*, **12,** 1–77.

KETTLEWELL, H. B. D. 1942. A survey of the insect *Panaxia* (*Callimorpha*) *dominula. Trans. S. Lond. ent. nat. hist. soc.*, 1–49.

KETTLEWELL, H. B. D. 1958. A survey of the frequencies of *Biston betularia* (L.) (Lep.) and its melanic forms in Great Britain. *Heredity*, **12,** 51–72.

KLOMP, H. 1962. The influence of climate and weather on the mean density level, the fluctuations and the regulation of animal populations. *Arch. Néerl. Zool.*, **15,** 68–109.

KLOMP, H. 1966. The dynamics of a field population of the pine looper. *Adv. Ecol. Res.*, **3,** 207–305.

KLUIJVER, H. N. 1951. The population ecology of the Great Tit, *Parus m. major* L. *Ardea*, **39,** 1–135.

KOJIMA, K. I. and YARBROUGH, K. M. 1967. Frequency dependent selection at the esterase-6 locus in *Drosophlia melanogaster. Proc. nat. Acad. Sci. U.S.A.*, **57,** 645–49.

KOSTITZIN, V. A. 1939. *Mathematical Biology*. Harrap, London, Toronto, Bombay, and Sydney.

KREBS, C. J. 1966. Demographic changes in fluctuating populations of *Microtus californicus. Ecol. Monogr.*, **36,** 239–73.

KREBS, J. R. 1970. Regulation of Great Tit numbers. *J. Zool.*, **162,** 317–33.

LACK, D. 1944. Ecological aspects of species formation in passerine birds. *Ibis*, **86,** 260–86.

LACK, D. 1947. *Darwin's finches.* Cambridge University Press.

LACK, D. 1954. *The natural regulation of animal numbers*. Clarendon Press, Oxford

LACK, D. 1966. *Population studies of birds*. Clarendon Press, Oxford.

LAMB, H. H., LEWIS, R. P. W. and WOODROFFE, A. 1966. Atmospheric circulation and the main climatic variables between 8000 and O.B.C.: meterological evidence. pp. 174–217. in *World Climate from* 8000 *to* 0 *B.C.* Royal Meterological Society, London.

LAMOTTE, M. 1951. Recherches sur la structure genetique des populations naturelles de *Cepaea nemoralis*. *Bull. Biol. Fr. Belg.*, **35,** 1–238.

LANDAHL, H. D. 1955. A mathematical model for the temporal pattern of a population structure, with particular reference to the flour beetle. *Bull. math. Biophys.*, **17,** 63–77 and 131–40.

LANDOLT, E. 1957. Physiologische und ökologische Untersuchungen an Lemnaceen. *Ber. schweiz. bot. Ges.*, **67,** 271–410.

LEES, D. R. 1970. The *Medionigra* polymorphism of *Panaxia dominula* in 1969. *Heredity*, **25,** 470–5.

LESLIE, P. H. 1945. On the use of matrices in certain population mathematics. *Biometrika*, **33,** 183–212.

LESLIE, P. H. 1948. Some further notes on the use of matrices in population mathematics. *Biometrika*, **35,** 213–45.

LESLIE, P. H. 1957. An analysis of the data for some experiments carried out by Gause with populations of the protozoa, *Paramecium aurelia* and *Paramecium caudatum*. *Biometrika*, **44,** 314–27.

LESLIE, P. H. 1958. A stochastic model for studying the properties of certain biological systems by numerical methods. *Biometrika*, **45,** 16–31.

LESLIE, P. H. and GOWER, J. C. 1960. The properties of a stochastic model for the predatory-prey type of interaction between two species. *Biometrika*, **47,** 219–34.

LEVINS, R. 1968. *Evolution in changing environments*. Princeton University Press.

LEWIS, E. G. 1942. On the generation and growth of a population. *Sankhya*, **6,** 93–6.

LEWONTIN, R. C. 1958. A general method for investigating the equilibrium of gene frequencies in a population. *Genetics*, **43,** 419–34.

LEWONTIN, R. C. 1966. On the measurement of relative variability. *Systematic Zoology*, **15,** 141–2.

LEWONTIN, R. C. 1968a (Ed.) *Population biology and evolution.* Syracuse University Press.

LEWONTIN, R. C. 1968b. The effect of differential viability on the population dynamics of t alleles in the house mouse. *Evol.*, **22,** 262–73.

LEWONTIN, R. C. and DUNN, L. C. 1960. The evolutionary dynamics of a polymorphism in the house mouse. *Genetics*, **45,** 702–22.

LEWONTIN, R. C. and HUBBY, J. L. 1966. A molecular approach to the study of genic heterozygosity in natural populations. II. Amount of variation and degree of heterozygosity in natural populations of *Drosophila pseudoobscura*. *Genetics*, **54,** 595–609.

LI, C. C. 1955a. *Population genetics*. University of Chicago Press, Chicago.

LI, C. C. 1955b. The stability of an equilibrium and the average fitness of a population. *Am. Nat.*, **89,** 281–95.

LI, C. C. 1967. Genetic equilibrium under selection. *Biometrics*, **23,** 397–484.

LLOYD, G. H. 1962. The distribution of squirrels in England and Wales, 1959. *J. Anim. Ecol.*, **31,** 157–65.

LOCKET, G. H. and MILLIDGE, A. F. 1951. *British spiders. Vol. 1.* Ray Society, London.

LOTKA, A. J. 1925. *Elements of physical biology*. Williams, Baltimore (reprinted 1956 as *Elements of Mathematical Biology*, Dover Publications, New York).

LOTKA, A. J. 1945. Population analysis as a chapter in the mathematical theory of evolution. In. : *Essays on growth and form presented to D'Arcy Wentworth*

Thompson. Ed. Le Gross Clarke, W. E., Medawar, P. B. pp. 355–85. Clarendon Press, Oxford.

LUCAS, C. E. 1961. On the significance of external metabolites in ecology. *Symp. Soc. exp. Biol.*, **15,** 190–206.

LUND, J. W. G. 1965. The ecology of the freshwater phytoplankton. *Biol. Rev.*, **40,** 231–93.

MACARTHUR, R. 1955. Fluctuations of animal populations, and a measure of community stability. *Ecol.*, **36,** 533–6.

MACARTHUR, R. 1968. The theory of the niche. In: Lewontin (1968a), 159–76.

MACARTHUR, R. H. and CONNELL, J. H. 1966. *The biology of populations.* Wiley, New York, London and Sydney.

MACARTHUR, R. H. and WILSON, E. O. 1967. *The theory of island biogeography*. Princeton University Press, Princeton.

MAELZER, D. A. 1970. The regression of log N_{n+1} on log N_n as a test of density dependence: an exercise with computer-constructed density-independent populations, *Ecology* **51,** 810–22.

MALEK, I. and FENCL, Z. 1966. *Theoretical and methodological basis of continuous culture of microorganisms.* Academic Press, New York and London.

MANWELL, C. and BAKER, C. M. A. 1970. *Molecular biology and the origin of species: heterosis, protein polymorphism and animal breeding.* Sidgwick and Jackson, London.

MAYNARD SMITH, J. 1968. Evolution in sexual and asexual populations. *Am. Nat.*, **102,** 469–73.

MAYR, E. 1945. Some evidence in favour of a recent date. Symposium on age of distribution pattern of the gene arrangements in *Drosophila pseudoobscura. Lloydia*, **8,** 69–83.

MAYR, E. 1947. Ecological factors in speciation. *Evolution*, **1,** 263–8.

MAYR, 1951. Speciation in birds. *Proc. Xth Intern. Ornithol. Congr., Uppsala*, 1950, 91–131.

MAYR, E. 1963. *Animal species and evolution.* Harvard University Press.

MEDAWAR, P. B. 1960. *The future of man.* Methuen, London.

MILKMAN, R. D. 1967. Heterosis as a major cause of heterozygosity in nature. *Genetics*, **55,** 493–5.

MILNE, A. 1961. Definition of competition among animals. *Symp. Soc. exp. Biol.*, **15,** 40–61.

MONOD, J. 1950. La technique de culture continue; théorie et applications. *Ann. Inst. Pasteur*, **79,** 390–410.

MORAN, P. A. P. 1949. The statistical analysis of the sunspot and Lynx cycles. *J. Anim. Ecol.*, **18,** 115–16.

MORAN, P. A. P. 1950. Some remarks on animal population dynamics. *Biometrics*, **6,** 250–8.

MORAN, P. A. P. 1952. The statistical analysis of game-bird records. *J. Anim. Ecol.*, **21,** 154–8.

MORAN, P. A. P. 1953. The statistical analysis of the Canadian lynx cycle. I. Structure and prediction. *Aust. J. Zool.*, **1,** 163–73.

MORAN, P. A. P. 1954. The logic of the mathematical theory of animal populations. *J. Wildlife Management*, **18,** 60–6.

MORAN, P. A. P. 1962. *The statistical processes of evolutionary theory.* Oxford University Press, London.

MORAN, P. A. P. 1967. Unsolved problems in evolutionary theory. *Proc. 5th Berkley Symp. Math. Stat. Prob.*, **4,** 457–80.

MORAN, P. A. P. 1970. Haldane's dilemma and the rate of evolution. *Ann. Hum. Genet., Lond.*, **33,** 245–9.

MOREAU, R. E. 1948. Ecological isolation in a rich tropical avifauna. *J. Anim. Ecol.*, **17**, 113–26.

MORRIS, R. F. 1963. Predictive population equations based on key factors. *Mem. ent. soc. Can.*, **32**, 16–21.

NICHOLSON, A. J. 1932. The balance of animal populations. *J. Anim. Ecol.*, **2**, 132–78.

NICHOLSON, A. J. and BAILEY, V. A. 1935. The balance of animal populations. Part I. *Proc. zool. Soc. Lond.*, 551–98.

NORRINGTON-DAVIES, J. 1967. Diallel analysis of competition between grass species. *J. agric. Sci., Camb.*, **71**, 223–31.

NOVICK, A. and SZILARD, L. 1950. Experiments with the chemostat on spontaneous mutations of bacteria. *Proc. Nat. Acad. Sci.* USA. **36**, 708–19.

ODUM, E. P. 1953. *Fundamentals of ecology*. 1st ed. Saunders, Philadelphia and New York.

OHBA, S. 1967. Chromosomal polymorphism and capacity for increase under near optimal conditions. *Heredity*, **22**, 169–85.

OLDHAM, C. 1929. *Cepaea hortensis* (Müller) and *Arianta arbustorum* (L.) on blown sand. *Proc. malac. Soc. Lond.*, **18**, 144–6.

OLDROYD, H. 1969. Diptera brachycera Section (a) Tabanoidea and Asiloidea. *R. Ent. Soc. Lond. Handb. Ident. Br. Insects*, **9** (4), 1–132.

PARK, T., LESLIE, P. H. and MERTZ, D. B. 1964. Genetic strains and competition in populations of Tribolium. *Physiol. Zool.*, **37**, 97–161.

PARK, T., MERTZ, D. B., GRODZINSKY, W. and PRUS, T. 1964. Cannibalistic predation in populations of flour beetles. *Physiol. Zool.*, **37**, 289–321.

PENNYCUICK, L. 1969. A computer model of the Oxford Great Tit population. *J. theor. Biol.*, **22**, 381–400.

PIELOU, E. C. 1969. *An introduction to mathematical ecology*. Wiley-Interscience, New York.

PONTIN, A. J. 1961. Population stabilization and competition between the ants *Lasius flavus* (F.) and *L.niger* (L.) *J. Anim. Ecol.*, **30**, 47–54.

PONTIN, A. J. 1963. Further considerations of competition and the ecology of the ants *Lasius flavus* (F.) and *L.niger* (L.) *J. Anim. Ecol.*, **32**, 565–74.

PONTIN, A. J. 1969. Experimental transplantation of nest-mounds of the ant *Lasius flavus* (F.) in a habitat containing also *L.niger* (L.) and *Myrmica scabrinodis* Nyl. *J. Anim. Ecol.*, **38**, 747–54.

POSTGATE, J. R. 1965. Continuous culture: attitudes and myths. *Lab. Practice.* 1140–4.

POWELL, E. O. 1958. Criteria for the growth of contaminants and mutants in continuous culture. *J. gen. Microbiol.*, **18**, 259–68.

PUTWAIN, P. D. and HARPER, J. L. 1970. Studies on the dynamics of plant populations. III. The influence of associated species on populations of *Rumex acetosa* L. and *R. acetosella* in grassland. *J. Ecol.*, **58**, 251–64.

PUTWAIN, P. D., MACHIN, D. and HARPER, J. L. 1968. Studies in the dynamics of plant populations. II. Components and regulation of a natural population of *Rumex acetosella* (L.) *J. Ecol.*, **56**, 421–31.

RADOVANOVIĆ, M. 1959. Zum Problem der Speziation bei Inseleidechsen, *Zool. Jb.* (*Syst.*), **86**, 395–436.

RENDEL, J. M. 1968. Genetic control of developmental processes. In: Lewontin (1968a), 47–66.

REYNOLDSON, T. B. 1948. An ecological study of the enchytraeid worm population of sewage bacteria beds: synthesis of field and laboratory data. *J. Anim. Ecol.*, **17**, 27–38.

REYNOLDSON, T. B. 1966. The distribution and abundance of lake-dwelling Triclads —towards a hypothesis. *Adv. Ecol. Res.* **3**, 1–71.

REYNOLDSON, T. B. and BELLAMY, L. S. 1971. The establishment of interspecific competition in field populations with an example of competition in action between *Polycelis nigra* and *P. tenuis*. In: Den Boer and Gradwell, 1971.

REYNOLDSON, T. B. and DAVIES, R. W. 1970. Food niche and co-existance in lake-dwelling triclads. *J. Anim. Ecol.*, **39**, 599–617.

RICHDALE, L. E. 1957. *A population study of penguins*. Oxford University Press.

RICKER, W. E. 1954a. Effects of compensatory mortality upon population abundance. *J. Wildlife Management*, **18**, 45–51.

RICKER, W. E. 1954b. Stock and recruitment. *J. Fish. Res. Bd. Can.* **11**, 559–623.

RICKER, W. E. 1958. Handbook of computations for biological statistics of fish populations. *Bull. Fish. Res. Bd. Can.*, **119**, 1–300.

RILEY, G. A. 1963. Theory of food-chain relations in the ocean. In: *The Sea*, Vol. 2, Ed. M. N. Hill, Wiley, New York, London and Sydney.

ROBINSON, G. A. 1970. Continuous plankton records: variation in the seasonal cycle of phytoplankton in the North Atlantic. *Bull. mar. Ecol.*, **6** (9), 333–45.

ROSENZWEIG, M. and MACARTHUR, R. 1963. Graphical representation and stability conditions of predator-prey interactions. *Am. Nat.*, **97**, 209–23.

SALISBURY, E. J. 1942. *The reproductive capacity of plants*. Bell, London.

SANG, J. H. 1950. Population growth in Drosophila cultures. *Biol. Rev.*, **25**, 188–219.

SCHOENER, T. W. 1965. The evolution of bill size differences among sympatric congeneric species of birds. *Evol.*, **19**, 189–213.

SCHWERDTFEGER, F. 1941. Über die Uraschen des Massenwechsels der Insekten. *Z. angew. Entom.*, **28**, 254–303.

SEARLE, S. R. 1966. *Matrix algebra for the biological sciences*. Wiley, NY, London and Sydney.

SEATON, A. P. C. and ANTONOVICS, J. 1967. Population inter-relationships. I. Evolution in mixtures of *Drosophila* mutants. *Heredity*, **22**, 19–33.

SHEPPARD, P. M. 1952. A note on non-random mating in the moth *Panaxia deminula*. (L.) *Heredity*, **6**, 239–41.

SHEPPARD, P. M. 1953. Polymorphism and population studies. *Symp. Soc. exp. Biol.*, **7**, 274–89.

SHEPPARD, P. M. and COOK, L. M. 1962. The manifold effect of the Medionigra gene of the moth *Panaxia dominula* and the maintenance of a polymorphism. *Heredity*, **17**, 415–26.

SHORTEN, M. 1954. *Squirrels*. Collins, London.

SKELLAM, J. G. 1951. Random dispersal in theoretical populations. *Biometrika*, **38**, 196–218.

SLATER, P. 1958. The general relationship between test factors and person factors: application to preference matrices. *Nature, Lond.*, **181**, 1225–6.

SMITH, C. A. B. 1969. *Biomathematics. Vol.* 2. Griffin and Co. London.

SMITH, F. E. 1952. Experimental methods in population dynamics: A critique. *Ecol.*, **33**, 441–50.

SMITH, F. E. 1961. Density dependence in the Australian thrips. *Ecol.*, **42**, 403–7.

SMITH, F. E. 1963. Population dynamics in *Daphnia magna* and a new model for population growth. *Ecol.*, **44**, 651–63.

SOKAL, R. R. and SONLEITNER, F. J. 1968. The ecology of selection in hybrid populations of *Tribolium castaneum*. *Ecol. Mon.*, **38**, 345–79.

SOLOMON, M. E. 1962. Ecology of the flour mite, *Acarus siro* L. (= *Tyroglyphus farinae* De. G.) *Ann. appl. Biol.*, **50**, 178–84.

SONNEBORN, T. M. 1954. The relation of autogamy to senescence and rejuvenation in *Paramecium aurelia*. *J. Protozool.*, **1**, 38–53.

SOULÉ, M. and STEWART, B. R. 1970. The 'niche-variation' hypothesis a test and alternatives. *Am. Nat.*, **104**, 85–97.

SOUTHERN, H. N. (Ed) 1954. *Control of rats and mice. 3. House mice.* Clarendon Press, Oxford.

SOUTHERN, H. N. (Ed.) 1964. *The handbook of British mammals.* Blackwell, Oxford.

SOUTHERN, H. N. 1970. Ecology at the cross-roads. *J. Ecol.*, **58,** 1–11, *J. Anim. Ecol.*, **39,** 1–11, *J. appl. Ecol.*, **7,** 1–11.

SOUTHWOOD, T. R. E. 1966. *Ecological methods, with particular reference to insect populations.* Methuen, London.

SOUTHWOOD, T. R. E. 1967. The interpretation of population change. *J. Anim. Ecol.*, **36,** 519–29.

SPRENT, P. 1969. *Models in Regression.* Methuen, London.

ST. AMANT, J. L. S. 1970. The detection of regulation in animal populations, *Ecology*, **51,** 823–8.

TAMM, C. O. 1956. Further observations on the survival and flowering of some perennial herbs, I. *Oikos*, **7,** 273–92.

TEMPEST, D. W. 1970a. The place of continuous culture in microbial research. *Adv. microbial physiol.*, **4,** 223–50.

TEMPEST, D. W. 1970b. The continuous cultivation of microorganisms. 1. Theory of the chemostat. pp. 259–276 in *Methods in microbiology Vol. 2.* Ed J. R. Norris and D. W. Ribbons. Academic Press London and New York.

TURNER, J. R. G. and WILLIAMSON, M. H. 1968. Population size, natural selection and the genetic load. *Nature, Lond.*, **218,** 700.

USHER, M. B. 1966. A matrix approach to the management of renewable resources, with special reference to selection forests. *J. appl. Ecol.*, **3,** 355–67.

USHER, M. B. and WILLIAMSON, M. H. 1970. A deterministic matrix model for handling the birth, death, and migration processes of spatially distributed populations. *Biometrics*, **23,** 1–12.

VANE, F. R. 1961. Continuous plankton records: Contributions towards a plankton atlas of the north-eastern Atlantic and the North Sea. Part III. Gastropods. *Bull. Mar. Ecol.*, **5** (42), 98–101.

VAN VALEN, L. 1965. Morphological variation and width of ecological niche. *Am. Nat.*, **99,** 377–90.

VARLEY, G. C. 1947. The natural control of population balance in the knapweed gall-fly (*Urophora jaceana*). *J. Anim. Ecol.*, **16,** 139–87.

VARLEY, G. C. 1949. Population changes in German forest pests. *J. Anim. Ecol.*, **18,** 117–22.

VARLEY, G. C. 1957. Ecology as an experimental science. *J. Anim. Ecol.*, **26,** 251–61.

VARLEY, G. C. and GRADWELL, G. R. 1960. Key factors in population studies. *J. Anim. Ecol.*, **29,** 399–401.

VARLEY, G. C. and GRADWELL, G. R. 1962. The interpretation of insect population changes. *Proc. 15th Ann. Sess. Ceylon Ass. Adv. Sci.*, 142–156.

VARLEY, G. C. and GRADWELL, G. R. 1970. Recent advances in insect population dynamics. *Ann. Rev. Ent.*, **15,** 1–24.

WALLACE, B. 1970. *Genetic load.* Prentice-Hall, Englewood Cliffs, N.J., U.S.A.

WATT, K. E. F. 1959. A mathematical model for the effect of densities of attacked and attacking species on the number attacked. *Canad. Ent.*, **91,** 129–44.

WATT, K. E. F. 1964. Comments on fluctuations of animal population and measures of community stability. *Can. Ent.*, **96,** 1434–42.

WATT, K. E. F. 1965. Community stability and the strategy of biological control. *Can. Ent.*, **97,** 887–95.

WATT, K. E. F. 1968. *Ecology and resource management. A quantitative approach.* McGraw-Hill, New York.

WHITTAKER, R. H. 1970 *Communities and ecosystems.* Collier-Macmillan. London.

WILLIAMS, G. C. 1966. *Adaptation and natural selection. A critique of some evolutionary thought.* Princeton University, Princeton.

WILLIAMSON, M. H. 1949. A preliminary note on the ecology of Lycosids (Araneae). *Entomologists mon. Mag.*, **85**, 92–3.

WILLIAMSON, M. H. 1957. An elementary theory of interspecific competition. *Nature, Lond.* **180,** 422–5 and **181,** 1415.

WILLIAMSON, M. H. 1958. Selection, controlling factors and polymorphism. *Am. Nat.*, **92,** 329–35.

WILLIAMSON, M. H. 1959. Some extensions of the use of matrices in population theory. *Bull. math. Biophysics* **21,** 261–3.

WILLIAMSON, M. H. 1960. On the polymorphism of the moth *Panaxia dominula* (L.). *Heredity*, **15,** 139–51.

WILLIAMSON, M. H. 1961a. An ecological survey of a Scottish herring fishery. Part IV: Changes in the plankton during the period 1949 to 1959. *Bull. mar. Ecol.*, **5,** 207–23.

WILLIAMSON, M. H. 1961b. A method for studying the relation of plankton variations to hydrography. *Bull. mar. Ecol.*, **5,** 224–9.

WILLIAMSON, M. H. 1963. The relation of plankton to some parameters of the herring population of the north western North Sea. *Rapp. Cons. Explor. Mer.*, **154,** 179–85.

WILLIAMSON, M.H. 1967. Introducing students to the concepts of population dynamics. The Teaching of Ecology. Ed. J. M. Lambert. *Symp. Br. Ecol. Soc.*, **7,** 169–76.

WILLIAMSON, M. H. in press. The relation of Principal Component Analysis to the Analysis of Variance. *Int. J. Math. Educ. Sci. Technol.*

WILSON, E. O. 1965. The challenge from related species. pp. 7–27. In: *The genetics of colonizing species*, Ed. H. G. Baker and G. L. Stebbins, Academic Press, New York and London.

WOLDA, H. 1963. Natural populations of the polymorphic land snail *Cepaea nemoralis* (L.) *Arch. Néerl. Zool.*, **15,** 381–471.

WOLDA, H. 1969a. Fine distribution of morph frequencies in the snail *Cepaea nemoralis* near Groningen. *J. Anim. Ecol.*, **38,** 305–27.

WOLDA, H. 1969b. Genetics of polymorphism in the land snail, *Cepaea nemoralis. Genetica*, **40,** 475–502.

WRIGHT, S. 1955. Classification of the factors of evolution. *Cold Spring Harb. Symp. quant. Biol.*, **20,** 16–24.

YOUNG, S. S. Y. 1967. A proposition on the population dynamics of the sterile t alleles in the house mouse. *Evolution*, **21,** 190–2.

ZWÖLFER, H. 1971. The structure of parasite complexes and its effect upon populations of phytophagous host insects. In: Den Boer and Gradwell. 1971.

Index

cont.